IN SEARCH OF ANCIENT NEW ZEALAND

IN SEARCH OF ANCIENT NEW ZEALAND

Hamish Campbell & Gerard Hutching

Co-published with GNS Science, New Zealand

PENGUIN BOOKS

Published by the Penguin Group
Penguin Group (NZ), 67 Apollo Drive, Rosedale
North Shore 0632, New Zealand (a division of Pearson New Zealand Ltd)
Penguin Group (USA) Inc., 375 Hudson Street,
New York, New York 10014, USA
Penguin Group (Canada), 90 Eglinton Avenue East, Suite 700, Toronto,
Ontario, M4P 2Y3, Canada (a division of Pearson Penguin Canada Inc.)
Penguin Books Ltd, 80 Strand, London, WC2R 0RL, England
Penguin Ireland, 25 St Stephen's Green,
Dublin 2, Ireland (a division of Penguin Books Ltd)
Penguin Group (Australia), 250 Camberwell Road, Camberwell,
Victoria 3124, Australia (a division of Pearson Australia Group Pty Ltd)
Penguin Books India Pvt Ltd, 11, Community Centre,
Panchsheel Park, New Delhi – 110 017, India
Penguin Books (South Africa) (Pty) Ltd, 24 Sturdee Avenue,
Rosebank, Johannesburg 2196, South Africa

Penguin Books Ltd, Registered Offices: 80 Strand, London,
WC2R 0RL, England

First published by Penguin Group (NZ), 2007
3 5 7 9 10 8 6 4

Copyright © Institute of Geological and Nuclear Sciences Limited, 2007

The right of Hamish Campbell and Gerard Hutching to be identified
as the authors of this work in terms of section 96 of the Copyright Act
1994 is hereby asserted.

Designed by Seven
Prepress by Image Centre Ltd
Printed in China through Bookbuilders, Hong Kong

All rights reserved. Without limiting the rights under copyright
reserved above, no part of this publication may be reproduced, stored
in or introduced into a retrieval system, or transmitted, in any form
or by any means (electronic, mechanical, photocopying, recording
or otherwise), without the prior written permission of both the
copyright owner and the above publisher of this book.

ISBN: 978-014-302088-2

A catalogue record for this book is available
from the National Library of New Zealand.

www.penguin.co.nz

The formal search for ancient New Zealand had rudimentary beginnings in Wellington in the 1860s. This photograph is of Colonial Museum staff and admirers with a mounted whale skeleton. It is understood that the man sitting on the ground (to left) is Walter Mantell, son of Gideon Mantell (the English doctor who in the 1820s was the first person to recognise the existence of dinosaurs). James Hector (seated to the right) was the founding director of the New Zealand Geological Survey, now GNS Science, and the Colonial Museum, now Te Papa.

Pygmy Southern Right Whale (*Caperea marginata*); photograph taken in 1874 from near the original Colonial Museum, which was located at a site behind present-day Parliament House, Thorndon, central Wellington, with Tinakori Hill as a backdrop.

L–R, those seated are: Mantell (assistant director), William Skey (chemist), Hector (director), Mr Burton (taxidermist); those standing are: A. T. Bothamley (assistant curator), R. B. Gore (curator, administrator), T. W. Kirk (zoologist), Alexander McKay (geologist), H. S. Cox (geologist) and John Buchanan (botanist and draftsman).

The oldest greywacke rocks of New Zealand occur on the West Coast of the South Island. They are known to geologists as the Greenland Group and are of Ordovician age (490-443 million years old). They are most conspicuous on the coast between Greymouth and Westport such as here at Fourteen Mile Bluff.

This book is dedicated to New Zealand geologist and teacher John Douglas Campbell (1927–2001), known as Doug or 'JD'.

He was especially interested in the geological history of New Zealand and the Triassic in particular, the period between 251 and 200 million years ago. Raised and educated in Wanganui, he studied geology at Otago University in the 1940s under Professor Noel Benson, becoming a lecturer at Canterbury University (1951–1958) and subsequently at Otago University (1959–1992). He was interested in botany almost as much as geology.

For much of the last 10 years of his life he systematically explored the folded and slightly ripped pages of geological history recorded on the coast between Kaka Point and Nugget Point. Chasing an idea, he also walked along the edge of one particular chapter, known as the Otapirian Stage, from the Taringaturas in the west to Roaring Bay in the east.

He was trying to determine just how uniform and consistent the ancient Panthalassa ocean floor was at that time 205–200 million years ago, the last 5 million years of Triassic time. What he found was a remarkable uniformity: he encountered the same shelly fossils of brachiopods and molluscs over a distance of more than 150 kilometres. It was as if he had traversed a 150-kilometre length of sea floor parallel to the coast and in water depths of about 100 metres.

Yet this exercise did not require scuba gear, submarines or submersibles. All he needed was a good pair of legs and a dedicated driving companion to drop him off and pick him up as he traversed the grasslands and forests of the Taringatura, Hokonui and Kaihiku ranges, the upland sheep-grazing country of Southland, following the exposed rock ribs breaking through the pasture.

There are countless millions of sea-floor surfaces preserved in the sedimentary rocks that make up so much of the New Zealand land mass. Think of each one as a page and the fossils as the words. As our collective investigations continue, so does the accumulation of our knowledge. We are far from knowing every page, but we can be reasonably satisfied that we have sorted out the chapters.

In his quest to better understand what happened in latest Triassic time in Southland, Doug Campbell was reading the rocks. So, too, our search for ancient New Zealand: we must interrogate the rock record. Let us begin!

A handsome specimen of *Clavigera bisulcata* collected by Graeme Stevens from the western coast of the North Island north of Marakopa. The original calcite shell is preserved and it is 60 mm long and 40 mm high. This species is relatively common and is characteristic of latest Triassic rocks (Murihiku terrane) of New Zealand that are between 205 and 200 million years old (Otapirian Stage). It belongs to a distinctive group of fossils, the athyrids, that Doug Campbell took particular interest in. The number refers to the collection number within the National Paleontology Collection held by GNS Science.

Contents:

Preface   16

Part 1: Introduction
01/ Hello, Zealandia!   22
02/ The Earth's Crust   26
03/ Basalt, Granite and Volcanoes   38

Part 2: Gondwanaland (505–83 Million Years Ago)
04/ From Ur With Love: Our Oldest Stuff   58
05/ The Greywacke Story   64
06/ The Oldest Rocks: Western Province   74
07/ Hot Rocks: Median Batholith   84
08/ The Younger Old Rocks: Eastern Province   92

Part 3: Zealandia (83–23 Million Years Ago)
09/ Zealandian Dinosaurs   116
10/ Crust Busters: The Tyranny of Faults   130
11/ Life Aboard Zealandia   146
12/ The Immersion of Zealandia   160

Part 4: New Zealand (23–0 Million Years Ago)
13/ The Emergence of New Zealand   178
14/ The Rise of the Southern Alps (5–0 Million Years Ago)   196
15/ Riding the Past 120,000 Years   214

Picture credits   228
Index   230
Acknowledgements   236
The New Zealand Geological Timescale   238

These islands;
The remnant peaks of a lost continent,
roof of an old world, molten droppings
from earth's bowels, gone cold;
ribbed with rock, resisting sea's corrosion
for an age, and an age to come.

ARD Fairburn, from 'Dominion'

If New Zealand has a defining rock type, it would be greywacke. It is most easily observed on the coast such as here at Red Rocks Point on Cook Strait near Wellington. This view is looking west towards Sinclair Head. Greywacke forms more than 60% of the New Zealand landmass, including the Southern Alps and the main ranges of both islands. Greywacke is muddy grey sandstone derived from Gondwanaland, but in rare instances, such as here at Red Rocks, other rock types are present. Collectively, they reveal much about the ancient history of our landmass.

Preface

We cannot deny our origins, our whakapapa. We must explore them. We cannot help ourselves. It is part of being human. It involves asking questions and looking back through history and time.

This is precisely what geologists do, but, instead of sifting through written documents, they decipher the record as preserved in stone. This book is all about this fascinating enterprise and its application specifically to New Zealand.

Nearly 20 years have passed since the last comprehensive popular explanation of the origins of New Zealand. I am thinking of that wonderful book published in 1988 entitled *Prehistoric New Zealand*, written by Graeme Stevens, Matt McGlone and Beverley McCulloch, with Jack Grant-Mackie, Dallas Mildenhall and Richard Holdaway, and illustrated by Vivian Ward.

Since the advent of *Prehistoric New Zealand*, knowledge and understanding of New Zealand geology has increased substantially, and our scientific perceptions have changed in subtle but nevertheless significant ways. The methods we use as earth scientists to interrogate the rock record have also changed. No longer is a fossil shell simply a record of an organism that lived and died a long time ago. The shell itself preserves a record of the chemistry and temperature of the sea water that the organism grew in.

During the past 10 years the Institute of Geological and Nuclear Sciences Ltd (GNS Science) has substantially remapped New Zealand at a scale of 1:250,000. This huge undertaking is approaching completion and will see the whole country mapped on 21 map sheets. During the process many new fossil localities have been discovered, but only a few have turned our heads.

Perhaps the greatest developments have been in the field of tectonics, both present and past. Thanks to modern technology, we now have a much better grasp of crustal processes and their rates. Plate tectonic theory has matured into established knowledge. There have been significant advances in our understanding of the ancestry and ultimate origins or sediment sources of our older sedimentary and metamorphic rocks, from what are referred to as 'provenance studies'. We also have a better understanding of the biological ancestry and origins of the native New Zealand fauna and flora, not so much from the fossil record, but rather through the advent of modern molecular biology. Lastly, and most importantly, there has been a conceptual breakthrough in understanding the geological history of New Zealand. New Zealand is the emergent part of a sunken continent, Zealandia. The full significance of this is only just beginning to be realised. Recognition of Zealandia will have far-reaching implications for New Zealand, not just scientifically, but politically and economically. It will change our perception of what New Zealand is and who we are as a nation, especially in the determination of the extent of our Exclusive Economic Zone (EEZ). The nature of the crust is a significant factor in the definition of an EEZ.

This book, then, is an update, a report on progress. In some respects it is comforting and reassuring, and in other respects it is challenging and perhaps even disturbing. One thing is for sure: nothing is necessarily quite as it seems!

The book is also partly an attempt to stimulate fresh interest in the geological aspects of our world. However, it is primarily a narrative of the great moments in the geological history of New Zealand. It makes ample use of everyday analogies to help explain and demystify geology for all to understand and enjoy.

If you have ever looked out of your window and wondered how old those hills are, or how old the rocks that form them are, or where they came from and why they are there, or how we know these things in any case – then this book may help. Every rock tells a story, and this book is an attempt to tell some of the most important stories about New Zealand and its origins.

In Search of Ancient New Zealand is a voyage of discovery. It deals with the ongoing quest to reveal the ultimate origins of the New Zealand land mass, the antiquity of the land surface, and the ancestry of our modern plants and animals. It offers insights and holistic explanations of all that is unique to the New Zealand environment and its natural heritage.

The story begins with that defining moment in the history of New Zealand, the formation of the new continent of Zealandia. It then looks back in time and explores what is known of the content, history and assembly of Zealandia, and proceeds with the extraordinary story of the sinking of Zealandia. A substantive and fresh New Zealand emerges, only to be reshaped by the ravages of vigorous tectonism, the ice ages, and the arrival of man.

The story is largely geological, but is nevertheless holistic and embracing. Think of geology as a blended mix of applied physics, applied chemistry and applied biology. Soundly based on first principles and the laws of thermodynamics, geological knowledge relates to three common entities: rocks, fluids and time.

As technology has advanced, geology has enabled us to read the 'memory banks' of this planet with increasing ease. (By this, we refer to the record of natural events and processes that are systematically preserved in the rocks that make up the Earth's crust.) Of course, not everything is preserved or recorded, but it is astounding what is! The Earth's crust and its surface are the ultimate archives, the memory banks of our planet, which hold the answers to some of our biggest questions.

Some of New Zealand's oldest rock is this Takaka Marble, dated at between 450 and 443 million years old (latest Ordivician).

In the Beginning
Ranginui, the Sky Father, and Papatuanuku, the Earth Mother, lay together in a tight embrace. Then Tane, god of forests and birds, forced his parents apart to allow light to enter the world and the world to take shape.

Creating the Land
Hauling his net from the depths, the demigod Maui pulled up a huge fish from the sea, which became the North Island. His canoe became the South Island, and its anchor Stewart Island.

While he turned his back, his brothers hacked out their share of the fish of Maui; soon it became a mass of valleys and ranges, a rugged land.

Mountain Building
The central North Island mountains – Tongariro, Taranaki (Mt Egmont), Tauhara and Putauaki (Mt Edgecumbe) – once lived close together in harmony. But, in a fight for the affections of the maiden mountain Pihanga, Tongariro won, and the other mountains were forced to leave. Before the rising sun could fix them to the spot, they walked as far as they could. There they stand to this day.

The Ngai Tahu tribe tell this story of the creation of the Southern Alps (Ka Tiritiri o te Moana, the frothing waters of the ocean): when Aoraki and his brothers were voyaging in the South Pacific, their canoe was wrecked. Aoraki scrambled to the highest point of the upturned hull and was turned into stone, becoming the mountain known today as Aoraki (cloud piercer) or Mt Cook.

As he travelled across the Southern Alps, Rakaihautu carved out the lakes of the South Island.

Hot Spots
Feeling cold and dispirited as he explored the North Island, Ngatoroirangi called to his sisters, Te Hoata and Te Pupu. They came underground from Hawaiki in the form of fire. When they appeared above the ground, they formed the boiling mud-pools, volcanoes and geysers that are famous in this region.

The Earth Moves
In tradition, taniwha (water spirits or monsters) shaped many of New Zealand's rivers, lakes and harbours. At Wellington Harbour an earthquake stranded a taniwha above ground, so that he formed a stretch of land.

Part 01:

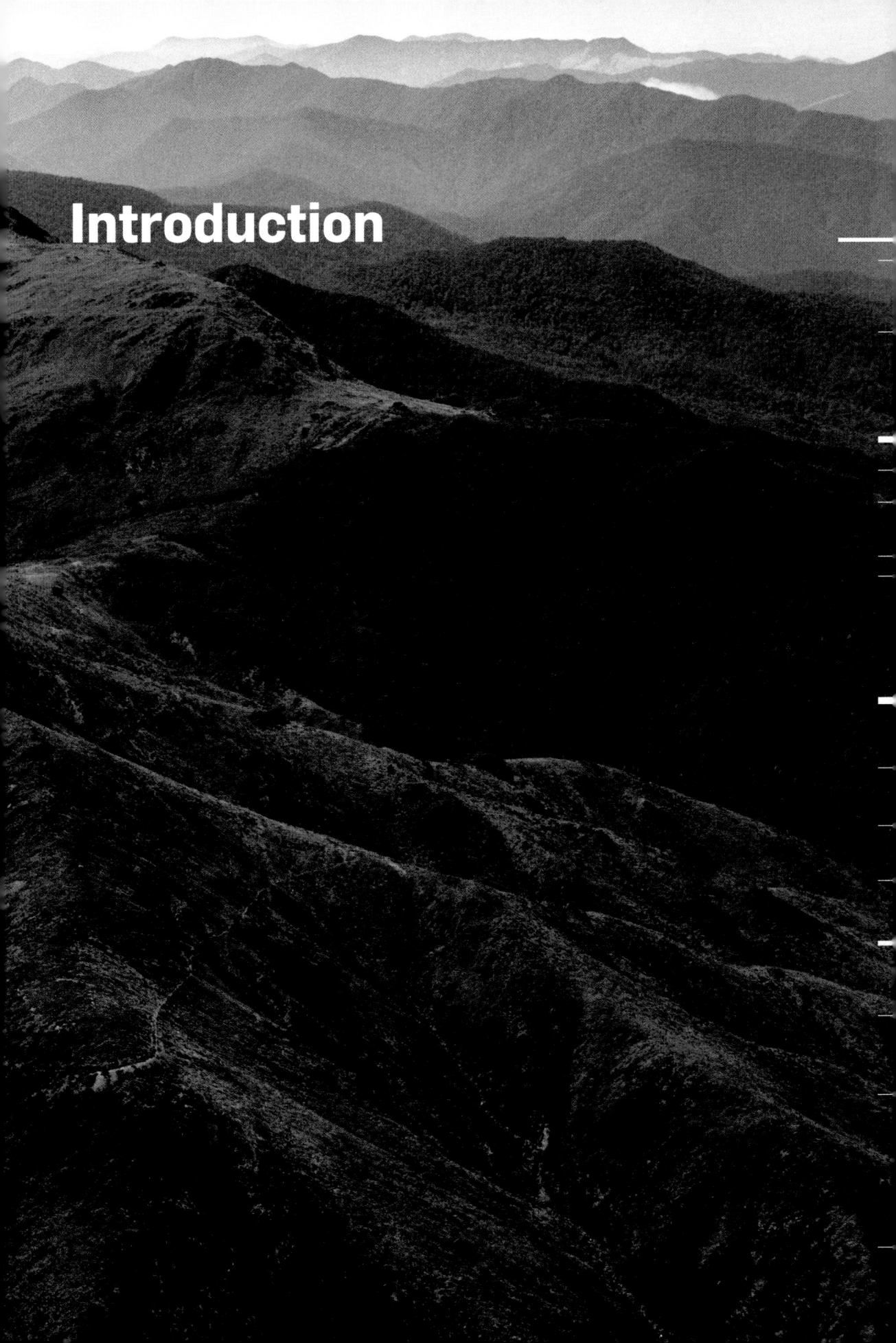

Introduction

01/ Hello, Zealandia!

Discovering Zealandia, The Eighth Continent

Zealandia is the name that geologists have given to the 'New Zealand continent', a vast geographic entity of momentous significance in world history. However, we have realised the scientific importance of its presence only within the past few decades. From a wider, popular perspective, it warrants greater appreciation and understanding.

Strictly speaking, a continent is defined as a large, discrete mass of land separated by expanses of water. Clearly, Zealandia does not conform to this definition, as it is largely underwater. However, geologists argue in favour of a revised definition of 'continent' as follows: an area of continental crust, irrespective of whether it is land or submarine. There are two such submerged continents: Zealandia (3,500,000 square kilometres in area) and the Kerguelen continent (1,131,000 square kilometres). If we accept this definition, then Zealandia and the Kerguelen Plateau join Asia, Africa, Antarctica, North America, South America, Australia and Europe as continents.

Captain Cook's Secret

In 1768 Lieutenant James Cook was chosen by the Royal Society of London as the most suitably qualified naval officer in the Royal Navy to lead a scientific expedition to the Pacific. He had outstanding skills as a navigator, but more particularly as an astronomer, and this is exactly what was required. Only a few months before he set off, Captain Samuel Wallis and the *Dolphin* had arrived home in England in May 1768

1. Captain James Cook sailed south from Tahiti in 1769 with secret instructions from the British Admiralty to find the Unknown Southern Continent ('Terra Australis Incognita'). Although disappointed in his quest, he should not have been, for New Zealand's islands are the emergent tips, only 7 percent, of a continental iceberg.

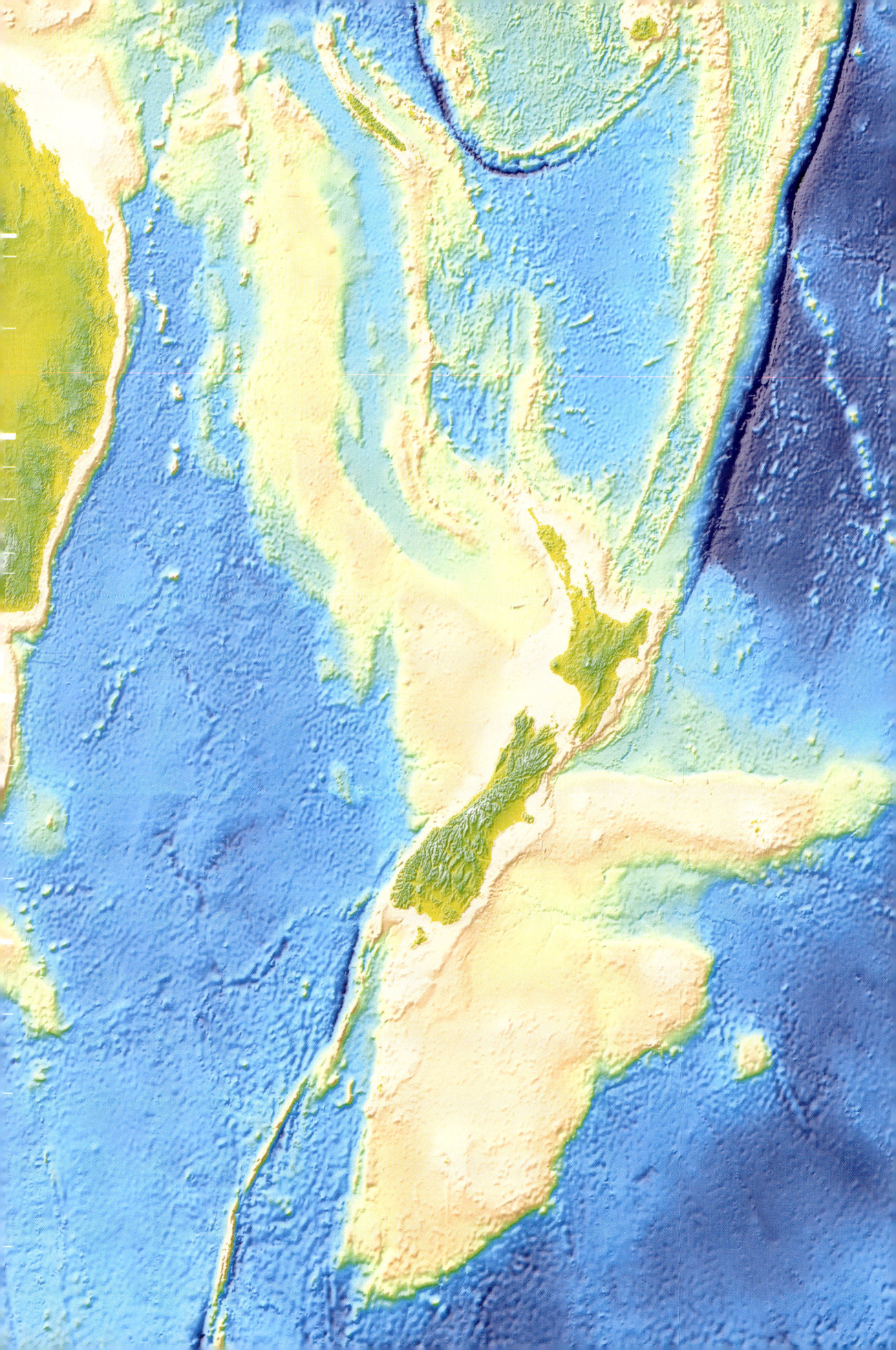

with news of the discovery of Tahiti. The timing was perfect! Tahiti was ideally located for observations of the transit of Venus.

Cook had express instructions to make astronomical observations of Venus as it crossed the face the Sun. Astronomers, including the Astronomer Royal of the day, Neville Maskelyne, had determined the exact timing of this event, which only happens four times every 335 years. It was going to happen on 8 June 1769. The reason for mounting a three-year excursion to make observations on a single day was simple: simultaneous observations of the transit of Venus from a number of different locations on Earth would enable scientists to calculate the distance between the Earth and the Sun, using simple triangulation. To know this would revolutionise navigation and cartography, with huge political and economic implications for humanity. Referred to as the 'Astronomical Unit', determining the distance between the Earth and the Sun would also transform astronomical understanding of the distances to the planets and stars, and the dimensions and enormity of space. Cook's observations would be part of an immense astronomical surveying exercise. As it happened, Cook was lucky: on the day, the Sun in Tahiti was cloud-free during the transit of Venus.

Rather than head home, Cook then headed south, following secret instructions to search for a great southern continent. The instructions were based on assertions from Captain Wallis, who had reported seeing land to the south of Tahiti. In reality, what he had observed were clouds.

Cook, of course, believed his search for the continent had failed. All he found was New Zealand, an elongate land area comprising a few relatively small islands. If only Cook were alive today! He would be amazed to learn that he had indeed found the great southern continent, but only the emergent part.

In order to understand New Zealand's new-found identity as the emergent part of a sunken continent, it is first necessary to appreciate what is known about the Earth's crust, and how that knowledge was discovered.

2. Present-day New Zealand is the emergent tip of the continent Zealandia, half the size of Australia and stretching from New Caledonia in the north to well beyond Campbell Island in the south, and to the Chatham Islands in the east. Most of this 'eighth' continent lies underwater, although it was once all land. The submarine shape and extent of Zealandia is defined by the distribution of continental crust. As a crude proxy, this includes everything above the 2500 m isobath. This water depth serves to delineate denser, less buoyant oceanic crust (everything below 2500 m) from less dense and more buoyant continental crust (above 2500 m).

02/ The Earth's Crust

The Ocean Floor: A Late Discovery

In the mid-1960s geologists came to the astonishing realisation that the crust beneath the ocean floor is profoundly different from the land. After many years of exploration, mainly for military purposes, it was established that the ocean floor is largely comprised of only one type of rock: basalt.

This was not and is not obvious. After all, the ocean floor is heavily disguised by a drape of camouflaging sediment (mud, silt, sand and gravel), not to mention being drowned by water. On average the ocean is 4–5 kilometres deep, and in places it exceeds 10 kilometres, such as in the Tonga–Kermadec Trench near New Zealand, and the Mariana Trench near the Philippines.

As a consequence of this discovery that the ocean floor is basalt, it slowly dawned on the geological world that the Earth's crust is characterised by two distinct components: continental crust and oceanic crust.

Continental Crust: Just Like the Moon!

If you look at the Moon when it is full, you see areas defined by two distinct colours. One is bright white, and the other is a greyer, silver colour. In fact, what you observe is reflected light from the Sun, and the bright white area is a more reflective part of the Moon's surface than the grey area.

If the Earth were stripped of its atmosphere, water and sediment, it too would look like the Moon. Viewed from afar, it would appear dominated by two colours, one bright white, and the other silvery grey.

1. The Moon is composed of just two rock types: granite, which makes up 80 percent of its surface; and basalt, the remaining 20 percent. From Earth during a full Moon, the granite shows up as white; the basalt, grey.

What are these two areas of colour? For the Moon, the much more extensive white area is dominated by granite and the grey area is dominated by basalt. The distribution of these two colours is a direct reflection of the distribution of the major types of rock that form the surface of the Moon.

The same would be true of the Earth were it stripped naked and gazed at from a distance, areas dominated by granite would appear bright white and areas of basalt, grey.

Only 20 percent of the Moon's surface is basalt, with almost none on the dark side. The other 80 percent is granite. It is of a particular type known as anorthosite, and compared to Earth, the basalt happens to be enriched in the metal titanium. So there are some differences. However, this is hardly surprising, as the Moon formed from a major collision between the primordial Earth and a large asteroid some 4.25 billion years ago. This was not very long after the birth of our Solar System, including the Sun and Earth, some 4.53 billion years ago. The Earth has subsequently changed greatly since the formation of the Moon, whereas the considerably smaller Moon has changed much less.

In fact, the Moon has remained relatively unchanged for the past 3 billion years, apart from the impacts of meteorites. Much of the surface of the Moon is therefore extremely ancient – 3 billion years old – and indeed may be regarded as the oldest surface that any of us see during our time on Earth. The last basalt lava flows to form on the surface of the Moon erupted about 1 billion years ago. Amazing!

Not Cheese, But Milk and Cream

The scientific world is quite certain about the types of rock that make up the Moon. Six of the Apollo space missions of the 1970s recovered large samples of

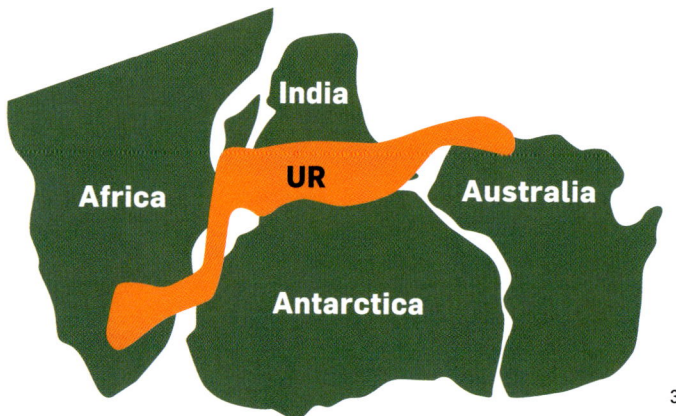

Moon rock (more than 350 kilograms in all), and from these much has been deduced about the Moon's composition, structure and behaviour, using basic physics and chemistry, experimental research, and modelling.

Some used to believe the Moon was made of cheese, but milk and cream would be a better, if still somewhat crude, dairy analogy. The Moon's metaphorical milk and cream is in fact basalt and granite, just like the Earth. We can think of basalt as acting in the same way as milk, and granite as cream. Both basalt and granite are igneous rocks and therefore start off as liquids, like milk and cream. The parallels continue, as basalt and granite are also derived from a common source, in this case the Earth's mantle.

Like milk and cream, basalt and granite vary in density, and they do so because they vary in composition. Basalt is denser because it is richer in iron- and magnesium-bearing minerals, while granite is less dense because it is richer in the lighter silicon- and aluminium-bearing minerals. Milk is denser than cream because it is richer in proteins, and cream is less dense than milk because it is richer in lipids. Cream floats on milk, and granite floats with respect to basalt. It is all a question of buoyancy.

Oceanic Milk and Continental Cream

If we think of oceanic crust as milk, and continental crust as cream, we are describing the Earth's crust in terms of two liquids. Of course, once at the Earth's

2. Image of a young girl, Rosa Ellingham, who is four-and-a-half years old. She is looking over a globe of the Earth that is a billion times older than herself at 4.53 billion years.

3. There was a time in Earth's long history when there were no continents. The first continent, Ur, formed around 3 billion years ago. Remnants of Ur have been recognised in western Australia, India, South Africa and Antarctica. Ur was followed by Arctica half a billion years later, then Baltica and Atlantica 2 billion years ago. The super-continent Rodinia was born 1 billion years ago.

5

surface, they become solid because they have cooled below their melting point and have crystallised. For basalt the melting point is about 1100°C, while for granite the point is much lower at about 850°C.

Cooling happens because the surface of the Earth loses heat extremely rapidly. The only protection between the Earth's surface and space is the hydrosphere (the oceans, lakes and rivers) and the atmosphere (air). When the air is clear at night, it is very cold. In fact, if space were not so incredibly cold, and rocks were not such poor conductors of heat, the Earth's surface would probably be a soft, hot, mushy substance and very difficult to walk on!

Ur: The First Continent

The milk and cream analogy neatly explains the distribution and behaviour of oceanic and continental crusts on Earth, including the phenomenon of 'sea-floor spreading' and continental drift. As time has passed, the Earth has evolved and the common liquid phases within the mantle have reached the surface as basalt and granite. From all available geological evidence, and in particular the systematic determination of the age of crystalline mineral formation in igneous rocks (basalt and granite) all over the Earth, it is quite evident that there has never been as much granite on the Earth's surface as there is now. Once at the surface of the planet, granite just floats around, like blobs of cream, changing shape and location. It is almost indestructible.

For this reason, the very first continent to have formed on the Earth's surface, named Ur, still exists. Granite differentiated from the mantle of the early Earth and appeared as a substantial surface area about 3 billion years ago. Remember that the definition of a continent is a surface area greater than 100,000 square kilometres. Ur conformed to this criterion. Despite subsequently breaking up and then amalgamating, it is still recognisable. The original blob of granite is now involved in four much bigger blobs of granite. Nowadays, parts of this first continent can be traced on four separate continents: India, Australia, Africa and Antarctica. Despite being shredded into pieces, the heritage and pedigree of Ur remain intact and known.

4. John Rogers, a well-known American geologist from North Carolina, USA, who has explored and championed the history of the continents more than most. Some of the wonderful names of previous continental configurations are his.

5. Geothermal hot springs such as the Champagne Pool at Waiotapu near Rotorua within the Taupo Volcanic Zone are natural laboratories. Research on extremophile organisms is providing insights into the primordial Earth, the origin of life and conditions on Ur, the first continent.

David Christoffel's Discovery

When Victoria University geophysicist David Christoffel told an earth sciences audience in 1970 that his measurements showed one side of the Southern Alps alpine fault had shifted 480 kilometres relative to the other, he was almost laughed out of the theatre.

At the time the theory of plate tectonics was freshly minted, hence the response from sceptics at the International Symposium on Crustal Movements. Much of the work published up to that date had been from the northern hemisphere; Christoffel was able to support the work with data he had been gathering from the southern hemisphere since the late 1950s.

Working with the New Zealand Navy, he designed and used magnetic sensing devices that were towed behind ships travelling between New Zealand and Antarctica. Although he did not realise the significance of the magnetic fluctuations at first, it became apparent, following Matthews, McKenzie and Vine's work, that these were measuring the rate of sea-floor spreading.

Christoffel's traced anomalies could be followed for long distances, and were parallel to the mid-oceanic ridge in the South-West Pacific. Even though they were never credited for it, Christoffel and colleague Robin Falconer also determined that Zealandia separated from Antarctica 83 million years ago.

Ur was just the beginning. Then came Arctica, followed by others. Continental crust formed ever-larger configurations of land, with wonderful names such as Baltica, Atlantica, Nena and Rodinia, 3–1 billion years ago. By 1 billion years ago, most of the Earth's continental crust had formed. Think of the continents as great blobs of cream (or as giant icebergs or puddles of oil on water), slowly moving and changing shape, coming together, colliding and pulling apart.

Humans have been mesmerised by the cream, the continents, the land. Only now can we see clearly, thanks to exploration of the oceans and modern technology.

Oceanic Crust: What of the Basalt?

What of the basalt that forms the ocean floor? In the mid-1960s it was determined that everywhere the sea floor is basalt. Even more extraordinary was the discovery that nowhere is the sea floor older than about 180 million years. This came as a rude shock and defied explanation. How could it be that the ocean floor was so much younger than the continents? The oldest known rocks on land were in excess of 3.8 billion years! Perhaps there was something wrong with the dating techniques? All aspects of radiometric dating were rigorously scrutinised and no fault in the theory or technology could be determined. It had to be true.

Our ability to determine how old rocks are has transformed geology and our understanding of the history of the Earth. It has enabled us to determine when geological events happened, as well as the rate at which geological processes occur. The implications of this kind of knowledge have been profound. We have since been able to determine the age of the Earth and our Solar System, and even shed light on the origin and significance of life itself.

Determining the age of igneous rocks such as basalt and granite is easy. A sample of rock is crushed so that particular minerals can be extracted. The minerals sought after are those containing either uranium or potassium. These are the most common radioactive elements and they are surprisingly abundant. Potassium is the eighth most abundant element in the crust, after oxygen, silicon, aluminium, iron, calcium, sodium and magnesium. For every tonne of crustal rock, about 20 kilograms is potassium. The most common potassium-bearing minerals are potassium feldspar, hornblende and mica (biotite, muscovite). Uranium is much less abundant than potassium at only 3 grams per tonne, but much greater than silver (0.7 grams) or gold (0.004 grams). The most common uranium-bearing minerals are zircon and monazite.

The Origin of Dating

How did we learn to date minerals and hence rocks? It really began with that extraordinary New Zealand

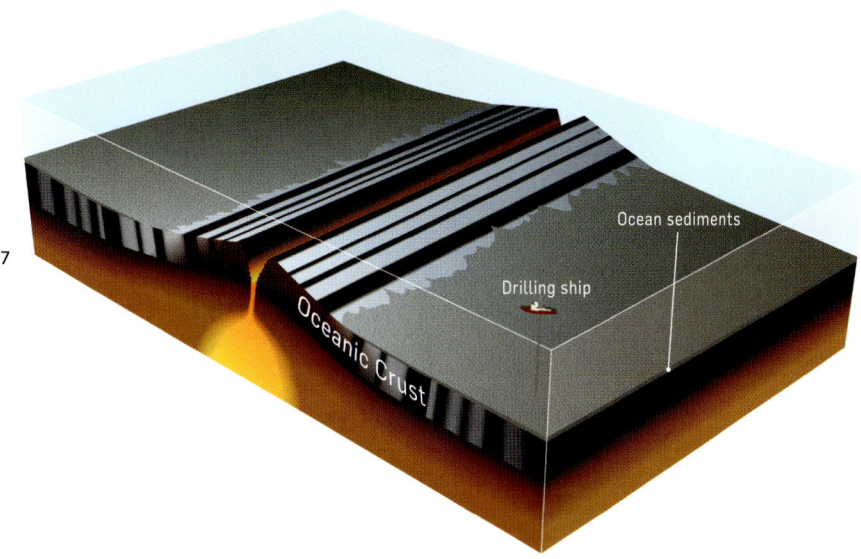

7

6

6. Drum Matthews, marine geophysicist and leading research scientist in the team that first stumbled upon the distinctive striped pattern of magnetic polarity preserved in sea floor basalts in the Atlantic Ocean.

7. A breakthrough in our knowledge of geology occurred in the 1960s when British and American scientists discovered that basaltic lavas erupted from submarine volcanoes along zones in the middle of oceans (mid-ocean ridges) record systematic changes in polarity of the Earth's magnetic field. The black 'stripes' in this diagram show times when the poles are normal – that is, when the North Pole is north and the South Pole is south. The white stripes represent times when the poles are reversed – that is, when north is south and south is north.

scientist, Ernest Rutherford. From his research on uranium and radioactivity, he realised that it should be possible to determine the age of a mineral. Knowing the rate of radioactive decay of uranium to lead, all that was necessary was a means of counting the relative proportions of uranium and lead atoms in the mineral being analysed. He published this idea with Frederick Soddy in 1903.

Rutherford developed a technique for separating and then counting atoms of different sizes. The concept was simple: it was like putting sheep through a race. The atoms (the sheep) were sent down a tube (a race); then the large atoms (uranium; the ewes) were deflected in one direction by a gate (a strong magnet) and the small atoms (lead; the lambs) were deflected in the other direction. The atoms were counted as they passed through the gate. This process is based on the fact that small atoms are drawn closer to the magnet than large atoms, rather like a stream of vehicles going around a bend: the smaller cars travel faster and hug the inside curve, while the large heavy trucks go much slower and track around the outer side of the bend.

The Magic of Magnetism

It took three young research scientists in Britain in the mid-1960s to figure it out: Drum Matthews, Dan McKenzie and Fred Vine. They were studying the history of magnetic polarity changes, as preserved in basalt on the sea floor in a small area of the Mid-Atlantic Ocean. Currently, the magnetic north pole is in its 'normal' position: at the North Pole. But through geological time, it has often been in the 'reverse' position: at the South Pole. This may sound confusing, but it is a common property of dynamos. Every now and then, polarity changes. Why it changes is another question, but the fact remains that it does.

Using a ship to tow a magnetometer, well clear of the magnetic influences of a metallic ship, these three young geologists recorded a curious stripe pattern of normal and reverse magnetic polarity, parallel to the mid-ocean ridge. What they observed was a symmetrical pattern of polarity change on either side of the Mid-Atlantic Ridge. It slowly dawned on them that the stripes represented elongate belts of sea-floor-forming basalt with an orderly systematic magnetic signature.

As liquid igneous rock cools and crystallises, iron-bearing minerals take on the characteristics, polarity and intensity of the ambient magnetic field that exists at the time. Rather like a supermarket bar code, the rock's secret code is locked in, but is easily readable with the right tools. In magnetic minerals, each iron atom behaves like a small compass that is randomly oriented while liquid, but

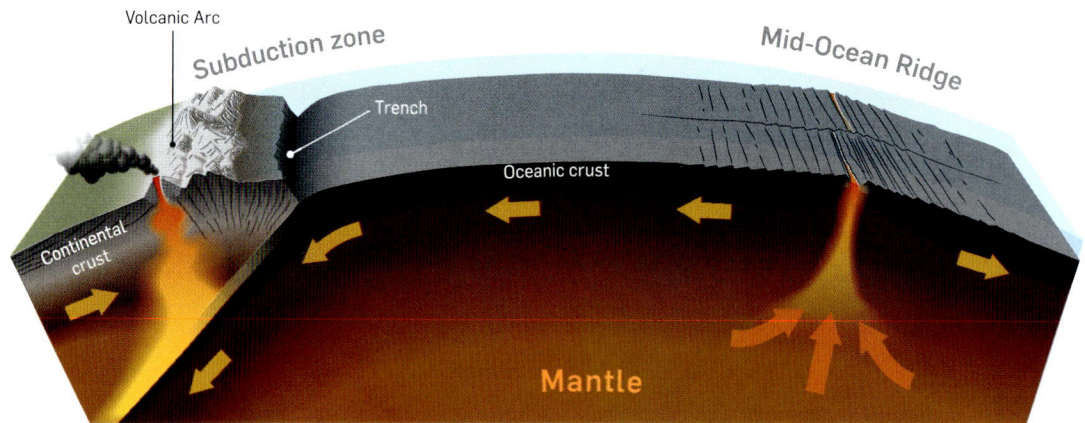

8. The Earth's geological cycle begins at mid-ocean ridges as fresh basalt lava erupting from submarine volcanoes. As if on a conveyer belt, or a sheet of cling film being drawn from its roll, the sea floor 'spreads' out taut to the subduction zone where plates collide. It is then drawn back down deep into the Earth. This is a slow process of endless recycling. Nowhere on Earth is the sea floor older than about 180 million years old.

sets instantaneously at the moment the liquid rock freezes and becomes solid. The iron atoms all align themselves in the direction of the North Pole at that very moment.

Once armed with age determinations of the basalts forming the Atlantic sea floor, the three young researchers were able to demonstrate satisfactorily the phenomenon of 'sea-floor spreading', and in so doing postulated the unifying theory that we know of as 'plate tectonics'.

What they had established is how the sea floor forms. Like milk, basalt flows. It rises as liquid rock (basaltic magma) along mid-ocean ridges and it descends along deep ocean trenches. Furthermore, the sea floor behaves like a giant conveyor belt that is driven in response to a gigantic convection system operating within the Earth's mantle.

They had satisfactorily established the answer to that riddle about the age of the ocean floor: it is young because it is mobile, and at any one time the ocean floor is no older than 180 million years. That is how long it takes to flow from the Mid-Pacific Ocean Ridge to the Mariana Trench, east of the Philippines. And it is precisely this flow that moves the granite about. Continental drift is a consequence of sea-floor spreading, the flowing basalt providing a mobile oceanic crust.

Origin of the Earth's Magnetic Field

The Earth behaves like a self-exciting dynamo, instantaneously and simultaneously producing an electrical field and a magnetic field. How does it do this? It is all to do with the innermost workings and structure of the Earth.

The inner core is solid and is comprised almost entirely of iron, whereas the outer core is liquid and is also enriched in iron. We know this from

9. The centre of the Earth is 6371 km from mean sea level; from here to 2900 km from mean sea level is the core, comprised of a solid inner and a liquid outer. The mantle stretches from 2900 km to an average of just 35 km, at which point the crust takes over: a relatively thin skin compared to what lies beneath. In fact, beneath the oceans the crust is only 7 km. The thickest continental crust is up to 80 km thick, beneath the Tibetan Plateau, and the Altiplano. The continental crust beneath Zealandia, and hence New Zealand, is unusually thin at less than 30 km, and this why 93 percent of Zealandia is submerged.

seismology (the study of earthquakes), experimental and theoretical physics, and chemistry. It is the relative motion between the inner and outer cores that creates the Earth's electromagnetic field, just like a self-exciting dynamo, which involves moving an iron bar inside a wire coil, or vice versa. And like a self-exciting dynamo, the polarity flips now and then.

The Mantle: Just a Skin

Today, this idea of a mobile crust to the Earth is old-hat, but in the mid-1960s it was all very difficult to accept. Perhaps the biggest stumbling block was the lack of a convincing mechanism to explain it all. After all, the world view was that the Earth's surface was fixed, except for localised effects of sediment accumulation, erosion, faulting and vulcanism. Nevertheless, plate

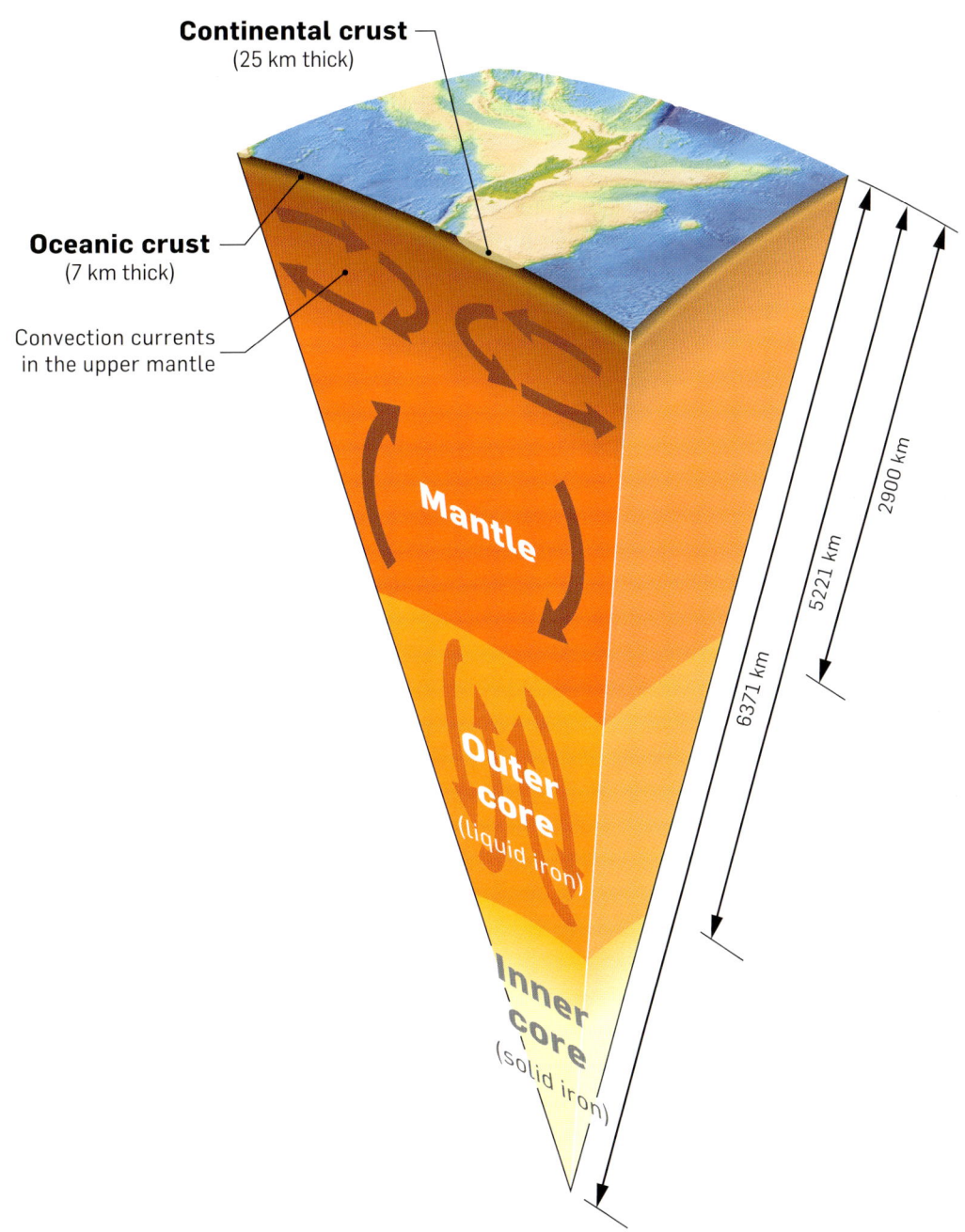

tectonic theory was so elegant and so powerful that it quickly became established as an holistic, all-embracing explanation of most geological phenomena on the surface of the Earth. The advent of plate tectonic theory revolutionised scientific understanding of the natural world.

We may now think of the Earth as a body, rather like our own bodies. The crust is just a skin and, like our skin, its structure and appearance is almost entirely determined by processes happening within the Earth, and in particular the mantle. The average thickness of oceanic crust is just 7 kilometres, and compared with the Earth's radius of almost 6400 kilometres, the crust is indeed thin and skin-like. Continental crust is thicker and attains maximum thickness of 80 kilometres below the highest areas on Earth, namely the Altiplano in South America and the Tibetan Plateau in central Asia.

A Skin of Frozen Liquids

If the crust is just the skin to the mantle, then the mantle assumes enormous importance. Understanding how the mantle works is the key to understanding how the crust works. As has often been said: 'The answer lies in the soil'. To understand how our skin forms and works, we need to understand how our bodies work.

To continue the skin analogy, the crust is an expansive surface area that behaves as a cooling surface between the hot mantle and space. In fact, if space were not so incredibly cold and rock were not such a poor conductor of heat, the crust would be hot and mushy, and uninhabitable. The crust may also be thought of as a frozen edge to the mantle: the igneous rocks that form the crust are frozen liquids (basalt and granite) that have been cooled below their respective melting temperatures.

In this regard, it is interesting to speculate on the origin and evolution of life of Earth: following the formation of our Solar System, life could have originated only once the temperature on the surface of the Earth was low enough, probably less than 100°C. It is interesting to speculate further that this important moment in Earth's history may not have been all that long after the planet formed. What we do know is that the next really big step, the appearance of complex multicellular life, occurred less than 1 billion years ago.

Mantle Rocks

From geophysical research, we know that the mantle is solid, not liquid. We have learned this from seismology; through simply observing the behaviour of earthquake waves as they pass through the Earth, we know that waves of seismic energy sweep through the mantle causing it to vibrate as if it were a gigantic gong. The mantle is as tough as steel. Yet despite this, it is hot and mobile.

The rock that makes up the upper mantle is peridotite, which is comprised largely of iron

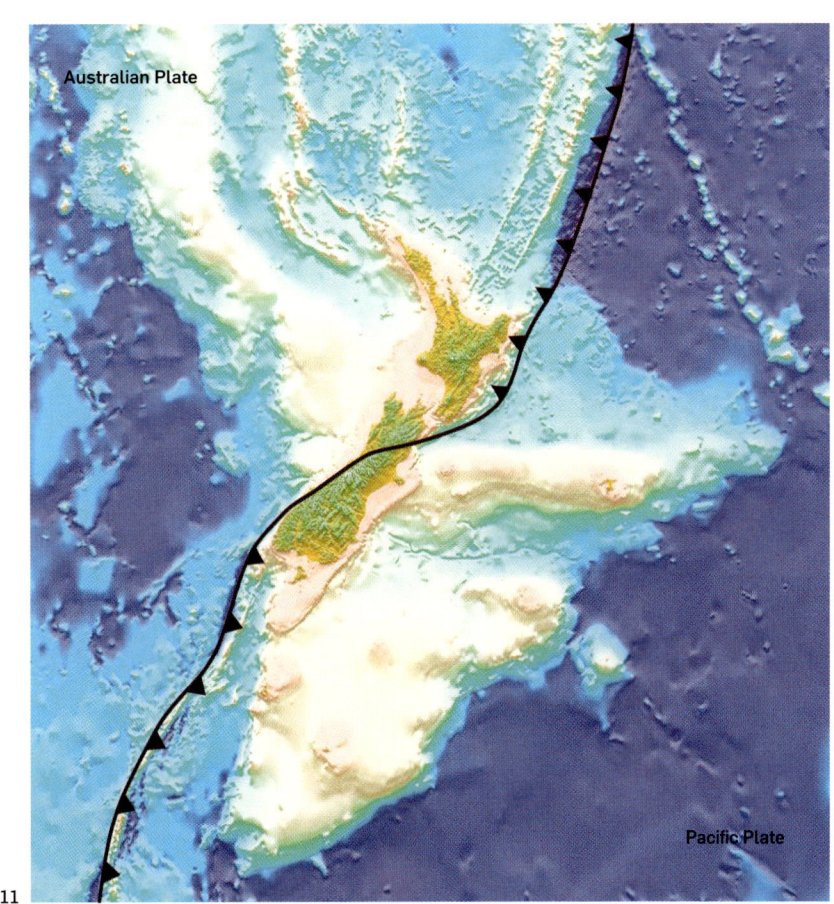

10. The Hope Fault is a dramatic feature cutting sharply through the Marlborough landscape to the north of Kaikoura in the northern part of the South Island. It is an active segment of the plate boundary between the Pacific Plate to the south-east (the lower-lying land on the right) and the Australian Plate to the north-west (the Seaward Kaikoura Mountains on the top centre and left). The collision between the plates is forcing the mountains up and at the same time, they are sliding past each other. The Pacific Plate is moving west and the Australian Plate is moving north.

11. When two worlds collide: this diagram shows the location of the modern plate boundary and how, in the northern end of New Zealand, the Pacific Plate is subducting (being drawn under) beneath the Australian Plate, but how to the south the Pacific Plate rides over the Australian Plate. Along the Southern Alps the plates slide past one another. The teeth point in the direction of subduction.

magnesium silicate minerals and is dense and heavy. Examples of peridotite (mantle rocks that have reached the surface by one means or another) are relatively common. Much less common, however, are samples of the minerals and rocks that make up the great bulk of the mantle and indeed must be regarded as the most common (yet unseen) material on Earth: perovskite. This mineral dominates the mantle for 660–2900 kilometres of the Earth's radius.

The Nuclear Source of Heat in the Earth

Temperature increases with depth within the crust by about 25°C per kilometre on average. The heat flow varies enormously from place to place on the Earth's surface, depending on crustal thickness and proximity to heat sources such as rift zones or volcanoes. For instance, in the Gulf of Thailand (a rift zone) the heat flow is about 60°C per kilometre, whereas in parts of western New Zealand, where there are thick sediment accumulations of up to 8 kilometres, it is only 15°C per kilometre.

Most of the heat generated within the Earth's mantle and crust is nuclear energy produced by the decay of the common radioactive elements, namely uranium, potassium and thorium. There is a probable second source of energy in the mantle, also nuclear, but produced within the Earth's core and considered to be from potassium.

New Zealand prides itself on being 'nuclear-free' and yet we have active volcanoes and geothermal fields. All the heat associated with these natural processes is nuclear energy. Furthermore, the driving force of plate collision through New Zealand is also natural thermonuclear energy. There are few places on Earth where the effects of large-scale nuclear power generated entirely within the Earth is so 'in your face'. How ironic! However, it needs to be said that New Zealand is free of man-made nuclear energy and its attendant environmental hazards and political risks.

Convection

The mantle acts as a kind of nuclear heat-exchange system – a heat pump. Heat rises and cools, and it is this convection process that is the driving force of sea-floor spreading and continental drift.

Heat convection is responsible for the movement of the oceanic crust. The same flow is responsible for the movement of the continental crust. In this regard, the Earth's crust is not unlike the skin that develops on porridge or viscous cooked fluids such as custard or milk. And just as in cooking, heat cells form with clearly defined boundaries between them, with zones of hot material rising and zones of cool material descending.

Since the mid-1960s, modern technology has enabled the measurement of these processes with remarkable precision. Computers, satellites and lasers have been the tools of preference and the results have been totally convincing. Direct measurement has effectively transformed plate tectonics from theory to fact, so that we know precisely how fast Christchurch and Dunedin are moving with respect to Wellington and Auckland.

Tectonic Plates

This thermal convection phenomenon explains why the Earth's crust is made up of a number of plates: their geometry is a reflection of the size and shape of the heat cell that is at work beneath them within the mantle. Tectonic plates are rigid entities, like plates of armour that are moving with respect to each other. There are about 15 plates. In New Zealand we are familiar with two of them: the Pacific and Australian plates. The New Zealand land mass straddles a segment of the collision boundary between these two plates.

Plates consist of some oceanic crust and some continental crust, but the relative proportions vary enormously. For instance, the Pacific Plate is gigantic. It is the largest plate of all, covering much of the Pacific Ocean and extending to the Americas. However, it is almost entirely comprised of oceanic crust. Surprisingly, by far the biggest area of continental crust on the Pacific Plate is that forming much of the South Island of New Zealand, the Chatham Rise and the Campbell Plateau.

In order to understand the relative motion involved and controlling the plate collision boundary through New Zealand, it is necessary to consider the motion not just of the immediate Pacific and Australian plates, but also that of the Antarctic Plate to the south. And as if this were not enough, the effects of all other plates need to be taken into account as well! After all, the plates are irregular-shaped, rigid, curved sheets that are all in relative motion with each other.

The plates are curved, which is why the Earth is a globe (not flat, as some people still believe). The plates are like the leather patches that when sewn up form a football, or the segments of peel from a peeled orange, or the large fragments of a broken eggshell. They fit together jigsaw-like, yet must conform to the shape of a sphere or globe. In fact, the Earth is not a perfect sphere, because the diameter at the equator is 27 kilometres greater than the diameter through the North and South poles. This is an inflation or expansion effect generated by the Earth's spin about its north–south-oriented axis.

Why Zealandia is So Wet

Much of this continental crust (the Chatham Rise and the Campbell Plateau) is below sea level. This is because it is not thick enough to be buoyed above sea level; the crust is simply too thin. Zealandia is only about 25 kilometres thick. Much of it lies 1–2 kilometres below sea level. Modern coloured bathymetric maps show the distribution of the oceans in terms of their water depth and depict the extent of Zealandia exactly. Until their advent, Zealandia was obscure, hidden and effectively unrealised and unknown. The seven better-known continents are 35–45 kilometres thick, and accordingly they stand 1–2 kilometres above sea level. It is all a question of buoyancy, and buoyancy is all about relative density.

A good analogy is ice on a pond. If it is thin, it will lie just beneath the water level. If it thickens, it will stand above the water level. It is the slight but nevertheless significant difference in density between water and ice that governs this phenomenon. Because ice is less dense than water, the thicker the ice, the higher it will sit. The same is true of continental crust: the thicker it is, the higher it rides with respect to the Earth's surface. It is less dense than oceanic crust, but only slightly so. Yet the significance of this difference is profound. It is the key to comprehending how the crust works, its structure and behaviour.

Zealandia Explained At Last

It is now possible to state how and why Zealandia formed and then sank in a way that is understandable. Because of an unknown and perhaps unknowable quirk of nature, thermal convection in the mantle suddenly changed or switched, and a new convection-cell geometry developed within the mantle beneath the eastern-most continental crust of the Gondwanaland supercontinent. A rift became established with considerable new granite emplaced. Eventually clean separation was achieved, creating a separate continental entity, Zealandia. The great blob of Gondwanaland had been reduced in size to produce a separate smaller blob. Fresh oceanic crust began to flow along a new mid-ocean ridge, giving rise to the Tasman Sea floor. The oldest basalt in the Tasman Sea floor formed 83 million years old, and therefore it can be claimed with some considerable certainty that Zealandia was born about 83 million years ago.

As it rifted away, Zealandia was stretched and thinned. The inevitable consequent loss of buoyancy resulted in slow sinking. Zealandia was to behave like a slowly sinking ship for 60 million years.

12. Buoyancy is the key to understanding how the Earth's crust works. The density contrast between water and ice is slight, yet sufficient for ice to float with respect to water. The same is true for oceanic and continental crust. Oceanic crust is denser than continental crust and therefore continental crust 'floats' with respect to oceanic crust. This explains continental drift. The continents are floating! Furthermore, the thicker the continental crust, the more buoyant it is and the higher it will rise, just as the ice does in these glasses of water.

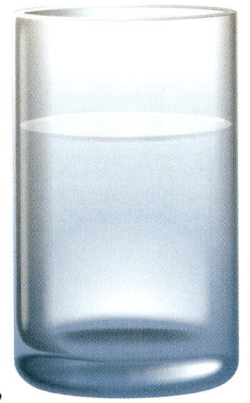

03/ Basalt, Granite and Volcanoes

What are Rocks and Minerals?

Rocks are solid materials comprised of minerals. All minerals are crystalline and all crystals grow in the presence of a fluid (liquid or gas). When crystals grow, minute spaces form between the crystals (interstices) and these spaces are filled with fluid. Rocks often have cracks and fractures that are also occupied by fluid. In this sense, although rocks may appear to be totally solid, in actual fact they invariably contain some fluid. The quantity varies enormously.

At least 3700 minerals have been named and described. Each has its own unique chemical composition and physical characteristics. They are combinations of the naturally occurring elements, of which there are just over a hundred. Yet there are only eight elements that form 90 percent of the minerals of the Earth's crust. These are oxygen, silicon, aluminium, iron, calcium, sodium, potassium and magnesium. The Earth may be thought of as 'the oxygen planet'. Almost 50 percent of an average rock is oxygen by weight, and about 90 percent oxygen by volume. The two most common elements are oxygen and silicon. It is hardly surprising, therefore, that one of the most common minerals is silicon dioxide or quartz (SiO_2).

Rock Types

There are some natural solids, such as obsidian and pumice (volcanic glasses with varying gas bubble content), which may be described as 'rock'. However, strictly speaking they are not crystalline and are therefore best thought of as superquenched liquids that have cooled so fast that they did not have time to grow crystals. Rocks come in a bewildering variety of colours and textures. They are described as 'fresh' (newly exposed to air) or 'weathered'. Weathered rocks are like tarnished brass or an old face. Their appearance has been affected by the ravages of exposure to the principal agents of decay – namely oxygen, radiation, mechanical and biological degradation. If it were possible to eliminate these agencies, rocks would stay fresh forever.

There are three broad categories of rock: igneous, sedimentary and metamorphic.

Igneous rocks are rocks that have crystallised or cooled from a hot magma. In a sense, they are frozen liquids. Basalt and granite are typical examples of igneous rocks. They separate from the Earth's mantle as distinct liquids. The faster they cool, the smaller the crystal size. Basalt cools rapidly below its melting temperature (or freezing temperature) of 1100–1250°C, and forms a rock with small crystals. Granite cools and crystallises more slowly below its much lower melting point of 850°C, forming much larger crystals.

Sedimentary rocks form from fragments of pre-existing rock, crystal or biogenic material (organically produced hard materials that include skeletal elements and wood) that have accumulated within a fluid (liquid or gas, water or air) as sediment. Examples include conglomerate, sandstone, siltstone, mudstone, limestone, chert and coal.

Metamorphic rocks are pre-existing rocks that have been altered or recrystallised in the presence of fluid in response to pressure and/or temperature changes. Any pre-existing rock can be metamorphosed, whether it is igneous, sedimentary or metamorphic. Think of metamorphic rocks as having been pressure-cooked. The temperature, pressure and time involved can vary enormously.

It is interesting to note that oil, gas and coal are commonly referred to as 'fossil fuels'. This is true in the sense that they are ultimately derived from fossils, the preserved organic remains of dead organisms. However, strictly speaking, they

1. Once sought after for its minerals, the fabled Red Hills area of north-west Otago is now a Wilderness Area and part of Mt Aspiring National Park. The remarkable iron and magnesium-rich rocks were uplifted millions of years ago and wrenched apart by the Alpine Fault, a portion of it (Dun Mountain) now 460 km to the north.

2A & 2B. These photographs, taken through a microscope, reveal thin sections of basalt from Rangitoto in the Hauraki Gulf near Auckland. The rock is lava that erupted only 600 years ago. A 'thin section' is a very thin slice of rock, parallel-sided and only 0.03 mm thick. When light is passed through the slice, the constituent minerals reveal themselves. Each mineral has its own unique properties of light transmission and absorption. Geologists depend on this kind of analysis (petrographic) of rocks. This is how we determine what a rock is made of and how it differs from others.

3. First detected in 1972, the Louisville seamount chain is a series of about 60 undersea volcanoes to the east and south of the Chatham Islands, and equal to the distance between Los Angeles and New York. It is a chain of spent cones, with only one active centre way to the east of the Chatham Islands. The cones are being transported away from the hot spot in a north-westwards direction by sea-floor spreading. The oldest is slowly subducting in the Tonga-Kermadec Trench.

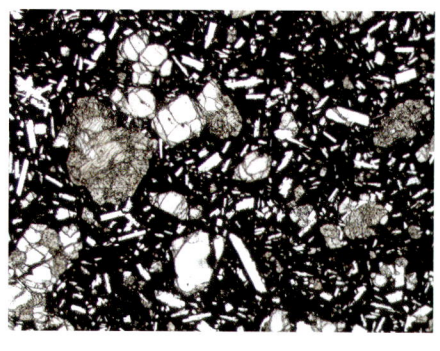

are the products of metamorphism. In this sense, oil and gas are best thought of as metamorphic fluids and coal as metamorphosed peat.

Peridotite

Peridotite is the rock that forms the Earth's mantle. There are a number of varieties of peridotite, and they all have wonderful names such as harzburgite, lherzolite and wehrlite that relate to the place names in Europe where they were originally described. The variety of peridotite depends on how much there is of the main minerals: calcium-rich feldspar, pyroxene and olivine. Surprisingly, one particular variety was named from New Zealand. This is 'dunite' and it is especially rich in olivine. Dunite was named by Ferdinand von Hochstetter after Mt Dun near Nelson.

Peridotite is a magnificent, dark green colour. Imagine the blue planet, Earth, skinned of its crust and devoid of its oceans and atmosphere with just the mantle showing: it would best be described as the green planet, and 'karaka green' at that.

Peridotite is the ultimate source of both basalt and granite. Both are generated as liquids that are distilled by the partial melting of peridotite. They form as separate entities because they have such different physical characteristics. In this regard, the milk and cream analogy we discussed in Chapter 2 works very well.

In order to appreciate how the crust works, it is imperative to understand the manifest characteristics, properties and behaviour of basalt and granite. In so doing, the geological history of New Zealand will make better sense and the search for ancient New Zealand will be much more comprehensible.

Basalt

Basalt is the most common rock on the surface of the Earth, but it is largely out of sight because it forms the oceanic crust beneath the sea floor and is therefore buried by sediment, not to mention water. It may also be thought of as the most common liquid phase within the Earth's mantle.

The most common mineral in basalt is calcium- and sodium-rich feldspar (plagioclase), making this the most common mineral in the Earth's crust. Other common minerals are pyroxene, olivine and iron oxides, but many other minerals can be present. Together these minerals produce rock that is greenish-grey to black in colour.

Most importantly, basalt is enriched in iron- and magnesium-bearing minerals and is therefore denser than granite, which is enriched in the lighter minerals silicon and aluminium.

Basalt either flows or is erupted onto the surface of the Earth in two ways.

Mid-ocean Ridges

First, basalt erupts along the mid-ocean ridges that are hot, up-welling zones of convection within the Earth, characterised by depressurisation of the Earth's mantle. This mechanism accounts for the basalt that forms oceanic crust, which is some 60 percent of the Earth's surface. It is responsible for

3

the Tasman Sea floor, not to mention those of the Pacific and Southern oceans.

The mantle is like a bottle of lemonade with the top on. Inside the bottle is a colourless liquid, but if the top is removed, zillions of gas bubbles appear out of nowhere. Where have all those bubbles come from? All you have done is take the top off, yet that is sufficient to alter the pressure of the liquid to allow the previously invisible gas phase to expand and escape. Even a slight pressure change can be enough to create a reaction.

In a sense, the same process is involved in producing liquid from peridotite, the rock that makes up the Earth's mantle. Confined under pressure, the basaltic liquid remains invisible, trapped in miniscule quantities between crystals. Breach the Earth's crust, and the mantle will be depressurised, allowing the basalt to flow like the bubbles of gas released when a lemonade bottle is opened.

Hot Spots

The second way that basalt erupts on the surface of the Earth is through hot spots. The sources of hot-spot vulcanism are so deeply rooted within the Earth that they are blind to the motion of the Earth's crust. They appear fixed within the mantle while the plates move over them. Yet they punch their way through. Accordingly, they leave a trail of spent volcanoes.

This process is rather like a sewing machine. Imagine sewing a sheet of fabric; a new stitch is made every 5 millimetres as the sheet is drawn through the machine. In the same way, a tectonic plate is drawn over a hot spot (the needle) and every few million years a new volcano (stitch) is made. An active volcano is immediately above a hot spot; all other volcanoes are dead, and the further away from the hot spot, the older the spent cone. Of course, the 'spot' may vary in size, as in Hawaii where there are several active volcanoes spread out over some distance.

Louisville Ridge

The nearest hot spot to New Zealand lies far to the east and south of the Chatham Islands. It has produced the Louisville Ridge, which is entirely submarine. Not one of the more than 60 volcanic cones is above sea level. In fact, the shallowest is still more than 400 metres deep. Amazingly, the trail of this hot spot is the mirror image of that producing the Emperor-Hawaii volcanic cones in the northern Pacific. This just goes to prove that sea-floor spreading of the Pacific Plate is indeed behaving as predicted according to plate tectonic theory: it is moving as a single entity, as if it were a sheet-like surface being dragged tautly as part of a vast revolving conveyor. It is being pulled up along the mid-ocean ridge and sucked down along ocean-margin trenches far to the west.

Auckland: Volcanic City
Auckland, New Zealand's largest and fastest-growing city, is built on a small, leaky hot spot. It is an active volcanic field, and in this respect Auckland is unique globally. There are plenty of cities built near active shield volcanoes, but no other city is built on an active basaltic volcanic field like that of Auckland. A shield volcano is broad, shield-shaped, just like Rangitoto Island, and usually large, like the volcanoes in Hawaii.

Auckland is studded by at least 49 basalt cones and craters that are thought to have all erupted within the past 60,000–140,000 years. So, on average there has been an eruption every 1500–3000 years. The most recent eruption was 600 years ago, producing Rangitoto Island, more voluminous than all the others

4

put together. There is no particular pattern to the distribution of Auckland's volcanoes, but there is a slight indication that the locus of eruption is moving to the north-east and that the eruptions are getting bigger.

There is no sign of activity at present, but nature is not regular and Auckland must therefore remain ever-vigilant. There is no reason to suppose that there may not be another eruption in the future. In the Auckland region, there are a number of hot water springs, evidence of geothermal activity lingering on, but not necessarily related to volcanic activity.

The Auckland volcanic field is being monitored by GNS GeoNet, so it is 'wired up', so to speak. We can expect distinctive seismic activity to precede an impending eruption as magma forces its way towards the surface, and fluid-related earthquakes have a very different signature from tectonic earthquakes.

Basalt, Granite and Volcanoes

Living in Auckland

People often ask where is the best place to live in Auckland – the answer is on a volcano! The theory goes like this: the small basalt plugs that form following an eruption behave like giant corks and are too strong for the next pulse of magma to cut through. It is much easier for the magma to find a new route around or between pre-existing basalt bodies.

Having said that, there is new geological evidence to suggest that at least some individual volcanoes within the field must have multiple-eruption histories. Detailed studies of drill-cores from several volcanic craters have revealed many more eruptions than previously thought. The Orakei Basin, for instance, has a record of at least 90 local Auckland eruptions within the past 90,000 years, but there are only 49 volcanoes. This implies that many of Auckland's volcanoes are not mono-genetic (meaning that they erupted only once), as previously thought.

In Auckland, the basalt volcanoes have punched through a succession of relatively weak sandstone that accumulated in deep water on the flanks of a submarine andesite volcano. This in turn is underlain by greywacke basement.

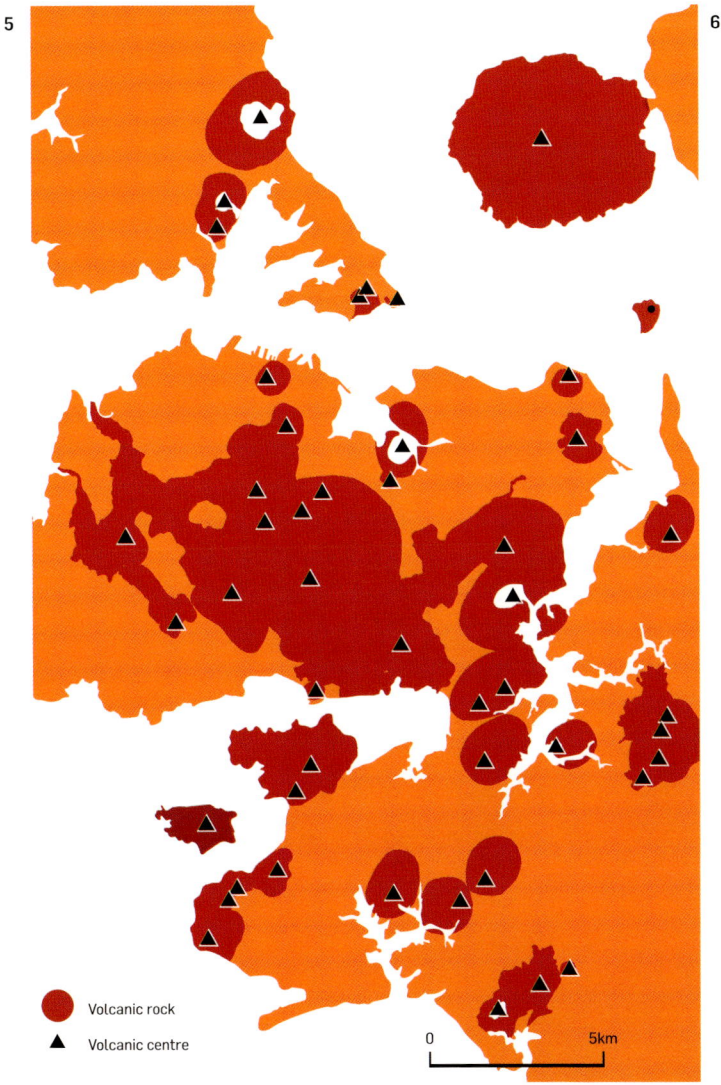

4. Known as the 'City of Sails', Auckland might just as accurately be described as the 'City of Volcanoes'. Lying astride a leaky hot spot, Auckland's gently sloping suburban hills disguise 49 craters and basalt cones. From the air, its volcanic past is more evident, as this aerial photo of Browns Island shows.

5. Auckland has 49 volcanoes in an area of 360 square km. A young field, its first volcanoes erupted between 60,000 and 140,000 years ago. It is expected to continue to be active for a million years.

6. The explosion crater that formed the Orakei Basin gives up its secrets. This core sample being examined by vulcanologist Graham Leonard is a record of numerous volcanic eruptions from the Auckland volcanic field and beyond.

7

Youngest Basalt in the North Island

A basalt eruption was responsible for New Zealand's most devastating historic eruption, the Tarawera Eruption of June 1886 that killed more than 150 people.

This eruption, sudden and unannounced, is thought to have been triggered by a chance injection of basalt magma into rhyolite magma. The two are very different brews in terms of both temperature and composition. Little wonder that the Earth went crazy. The eruption was short-lived, lasting less than 24 hours, but was very destructive. A livid vent developed over a length of 17 kilometres, causing extensive damage and most notably destroying the famous pink and white terraces, a major tourist attraction.

The terraces were a magnificent cascade of gently inclined, flat surfaces, with hot geothermal waters trickling over them. As the water flowed, it evaporated over the expansive surface area afforded by the terraces and, as a consequence, the silica precipitated out from the resulting supersaturated solution, forming the most intricate, filigree textures and patterns with exquisite colours. The finest silica growths were along the very edges of the terraces, where the evaporation rate would be at its highest. One set of terraces was brilliant white and the other, nearer Tarawera, was a stunning pink. The pink colour was due to tiny grains of a natural amalgam of gold.

The pink and white terraces were living, growing entities. What a shame that they should succumb to common old basalt! However, there is no reason why such surfaces could not develop again, given the right conditions.

There are younger lava flows in the North Island but these are of andesite composition rather than basalt. This is a subtle distinction perhaps, but important from a geological perspective. Composition of the lava conveys key information about the magmatic origins of a volcano and hence its tectonic significance. Andesite is associated with crustal processes, notably subduction, whereas basalt is more fundamental and is derived from the mantle. Subduction means 'to be led under', and specifically involves the sinking and drawing down (sucking) of oceanic crust beneath continental crust. The youngest lava flows in the North Island erupted from Ngauruhoe in 1975, when it last blew. Ngauruhoe is considered to be the youngest cone of the Tongariro volcano and is only about 2500 years old. A young nipper!

Our Old Basalt Volcanoes

The Chatham Islands, Auckland Island and Campbell Island are all remnants of large, extinct basaltic volcanoes. Dunedin is built within the breached centre of a large, basalt volcanic complex that erupted 16–10 million years ago. Otago Harbour is smack in the middle of the Dunedin volcanic complex and is surrounded by large, upstanding hills of basalt. Smaller, conical volcanoes within the greater Dunedin area stretch to the north as far as Palmerston and inland as far as Ranfurly.

Lyttelton is built in a similar large, basalt volcanic complex that forms Bank's Peninsula. This complex is younger than that of Dunedin and erupted 12–6 million years ago. Note that basaltic rocks in all of these old volcanoes include many different rock types including trachyte, basalt and phonolite, to name but a few.

Youngest Basalt in the South Island

Timaru is built around Caroline Bay, which is formed from a thick, extensive basalt lava flow that erupted from a source near Mt Horrible, 15 kilometres inland. This is the youngest lava flow in the South Island and is 2.1 million years old.

Young Basalts in Northland

There are numerous small basalt volcanoes in Northland, and some of them form beautiful cones and craters such as the one near Kaikohe, which is about 1 million years old.

7. In 1886 Tarawera erupted in a violent explosion, killing at least 150 people and creating a spectacular 17 km-long rent. This view is looking north-east along the rent, above Lake Tarawera to Tarawera and beyond to Mt Edgecumbe.

Flood Basalt

Given half a chance, basalt can pour out of the Earth's mantle almost uncontrollably, producing vast areas of 'flood basalt' known as 'large igneous provinces' (LIP, for short). Perhaps the best-known example of flood basalt is the Deccan Traps of southern India. There, a thick stack of lava flows forms a stepped topography. Surprisingly, it has been shown that this extraordinary volume of basalt was all extruded over a relatively short period of time, less than 1 million years, and was more or less coincident with the Cretaceous–Paleogene extinction event 65 million years ago that wiped out the dinosaurs. There has been some considerable debate about a possible causal relationship between these three events.

The Siberian Traps is another even larger LIP located in Russia and linked with the even more disastrous Permian–Triassic extinction event. And there are many other areas of flood basalt throughout the world.

New Zealand can boast two flood basalt associations. The first is Kirwan's Dolerite, which is of Jurassic age, and is situated on the West Coast of the South Island near Reefton. It is so distinctive that it can be identified as a remnant of a widespread eastern Gondwanaland LIP that is especially well known in Antarctica and Tasmania: the Ferrar Dolerite. This is the only certain remnant in New Zealand, but basalt dikes of Jurassic age that intrude granite on the Bounty Islands may also relate.

The second flood basalt association is the Hikurangi LIP, a submarine feature of unusually thick oceanic crust, occupying the large triangular area north of the Chatham Rise and adjacent to the coast stretching from Bank's Peninsula almost to Hawke's Bay. This is of Cretaceous age and is an area of anomalously thick oceanic crust. No wonder! We now know that it constitutes a LIP.

Granite

Granite is the most common rock on the continents. It is often referred to as a 'plutonic' rock, meaning that it crystallises at depth within the Earth's crust. Unlike basalt, it does not form on the Earth's surface. In this regard, it behaves very differently from basalt, and yet they are both igneous rocks that form from a molten state, a liquid. Whereas basalt is extruded onto the surface of the crust and cools quickly, granite is intruded within it and cools slowly.

This difference in the rates of cooling is why granite is comprised of large crystals and is 'coarse-grained', and basalt is comprised of small crystals and is 'fine-grained'. However, when granite magma reaches the Earth's surface, it cools quickly and produces a volcanic rather than a plutonic rock that is as fine-grained as basalt but is called rhyolite.

Granite is comprised of minerals that are rich in silicon and/or aluminium: quartz, potassium feldspar and mica. As with basalt, a variety of other minerals are present. The most significant of these less abundant minerals is zircon. This mineral plays a huge role in our understanding of geological history, for the simple reason that it is robust and can easily be dated.

Plutons and Batholiths

Granite forms as large, rounded, intrusive bodies of rock referred to as 'plutons'. A single mountain may constitute the remains of a single pluton. At a larger scale, involving whole mountain ranges or elongate belts that are tens to hundreds of kilometres long, the term 'batholith' is used. There are spectacular examples of plutons and batholiths in New Zealand, such as the Victoria Range near Reefton.

We know the age of granite plutons and batholiths from the dating of minerals, particularly zircons. But this is not the only mineral that can be dated in granite. There are quite a few others, including mica (biotite and muscovite), as well as potassium feldspar. In this sense, granite is an easily dated rock. It is richer than basalt in the common radioactive elements uranium and potassium, and accordingly has much more abundant uranium- and potassium-bearing minerals.

Four batholiths are recognised. The oldest is the Karamea Batholith 380–360 million years old (Devonian) and 3200 square kilometres in area. It forms a broad north–south tract of Westland. By far the most voluminous is the Median Batholith at

Basalt, Granite and Volcanoes

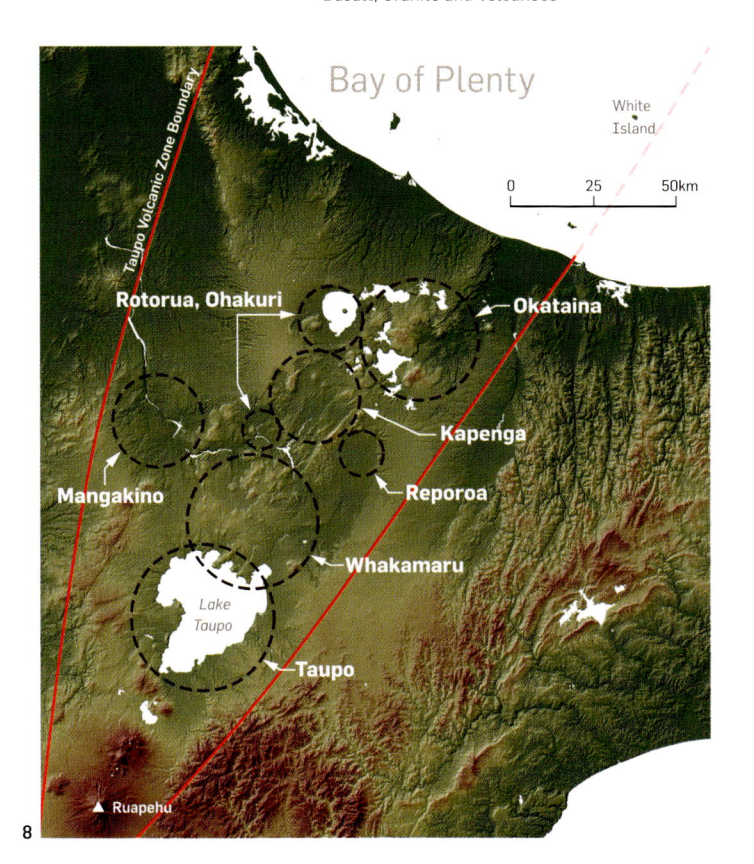

10,200 square kilometres in area. It forms a large part of Fiordland, much of Stewart Island, the Bounty Islands and a belt west of Nelson. It is also the longest-lived, ranging from 350 to 100 million years old. This represents Carboniferous to Cretaceous time. The other two are the Paparoa and Hohonu batholiths. These are comparatively small at 400 and 300 square kilometres, respectively, and are near Greymouth, also in Westland. They are both Cretaceous, 125–83 million years old.

The Median, Paparoa and Hohonu batholiths are of great significance in terms of the rifting of Zealandia from Gondwanaland. They were emplaced in the early phases of rifting, prior to separation.

8. The Taupo Volcanic Zone (defined by the red line) is a spectacular rift zone within continental crust. It runs from Tongariro National Park in the south to White Island in the north and contains eight calderas. It is one of the most geothermally active zones in the world and home to super-volcanoes comparable to Yellowstone in the USA. It is actively rifting open in an east–west direction.

Poignantly, the most voluminous pluton within the Median Batholith is named the Separation Point Pluton! This is superbly exposed in north-west Nelson between Golden Bay and Tasman Bay. It really does represent the onset of stretching and rifting of continental crust.

Greywacke, Son of Granite

Greywacke and schist (metamorphosed greywacke) constitute the most common and widespread hard basement rocks in New Zealand. They make up about 60 percent of our land mass and form the axial ranges of the North Island, including the Rimutaka, Tararua, Ruahine, Kaweka, Kaimanawa, Urewera and Hunua ranges. And in the South Island they form the Southern Alps, much of Westland, some of north-west Nelson, all of Marlborough, Canterbury, Otago and even northern Chatham Island. Only Stewart Island, Fiordland and Auckland Island are free of them.

Greywacke and schist are mentioned here because they are almost all derived from granite. Yet the mother granite batholith, from which the original sediment that makes up the greywacke and schist is derived, does not occur in New Zealand. It is clear from considerable research that all of it must have been derived from somewhere else within Gondwanaland. There is no suitable source in Zealandia or New Zealand. In this sense, the greywacke and schist may be thought of as recycled granite.

North Island Granites

Most granite in New Zealand is in the South Island, but there are two places in the North Island where granite is exposed: at Paritu on the north coast of Coromandel Peninsula; and the northern coast of Great Barrier Island. The Coromandel granite is 17 million years old, much younger than any granite in the South Island. It has been used widely for building purposes and is prominent in the Auckland Museum building in Auckland, and Parliament buildings in Wellington.

Drill-holes associated with the Wairakei geothermal field have penetrated

9. Lake Taupo, the sleeping giant. Peaceful though it appears today, New Zealand's largest lake was the scene of the world's largest volcanic eruption (the Oruanui) in recent history, 26,500 years ago, blanketing parts of the North Island with ash to a depth of 200 m.

10. Ash from the Oruanui eruption from Taupo Caldera blasted more than 1000 km across the ocean and is today preserved in peat on Chatham Island.

into the youngest granite known from New Zealand, only about 3 million years old.

It would be true to say that the northern part of central North Island is the centre of modern-day granite intrusion.

The Youngest Granite in New Zealand

The Taupo Volcanic Zone (TVZ) is an active rift within the continental crust of the North Island. It has been developing over the past 2 million years and is forming a slowly expanding V-shaped feature, with the open V represented by the Bay of Plenty, and Ruapehu marking the base of the V. According to satellite and laser surveying measurements, the TVZ is expanding in an east–west direction at 8–10 millimetres per year.

Now, just to add drama on a grand scale, the TVZ is the home of eight rhyolite calderas. Calderas are huge collapsed craters that have formed after extremely violent and productive volcanic eruptions. Because rhyolite lava (or granite magma) is viscous, it is very sticky, and when it erupts it tends to be extremely messy and violent. Accordingly, an eruption can be large and throw out vast volumes of volcanic matter. The result is a major depression that collapses in on itself.

This is exactly what Lake Taupo is. A beautiful scenic lake it may be, but the lake waters occupy a huge hole that has been created by multiple, violent rhyolite eruptions over the past 300,000 years. At least 28 eruptions of Taupo volcano have happened within the past 26,500 years. On average, this means that there is an eruption every 900 years, yet the most recent eruption happened about 1800 years ago. However, nature is not necessarily regular and the greatest recorded gap between eruptions is 5000 years. Nevertheless, the underlying process that is responsible for Taupo erupting is ongoing. We can be certain that it will erupt again, but we do not have answers to the two most critical questions: when will it happen; and how big it will be?

The Big One: Oruanui

The largest eruption we know of was 26,500 years ago and is known as the Oruanui Eruption. It was this event that largely determined the present shape of Lake Taupo. At the time it formed a caldera that was at least 500 metres deep and produced enough pumice, ash and rock to bury the entire North Island by 10 metres, if it were evenly spread out across the land, which it wasn't. The effects were nevertheless

10

immensely devastating and would have laid waste to vast tracts of land in the North Island. The impact on life would have been catastrophic and profoundly destructive.

We know all this from the detailed study of the nature of the material that erupted from the volcano (glass chemistry; rock, mineral, water and gas content; particle size; colour), as well as its stratigraphy (study of the layers or strata of rock, their nature and relationships), structure, age and distribution. Over the past three decades, many vulcanologists have laboriously interrogated the immediate Taupo landscape as well as the wider New Zealand land surface, and have accumulated a substantial knowledge of what happens when Taupo erupts. They have systematically traced the extent of individual sheets of ash and pumice. In so doing, they have given them names and have described them formally in the scientific literature. Many careers have been built on this kind of research, and some have become illustrious, globe-trotting vulcanologists, none more so than Bruce Houghton (University of Hawaii) and Colin Wilson (University of Auckland). The results of their research are fairly alarming: Taupo is now considered to be the most frequently active and the most productive rhyolite volcano in the world.

The Oruanui Eruption is beautifully recorded about 1000 kilometres from Taupo, in the Chatham Islands. Thick ash draped the landscape, leaving a compacted layer up to 18 centimetres thick. Yet this is not the only ash layer preserved in the Chatham Islands landscape. Thicker and older ash layers are present, and in particular a layer that is almost 50 centimetres thick, which is thought to be the Rangitawa Ash, erupted from the TVZ, probably Okataina Caldera, about 336,000 years ago. Volcanic ash is a general term for the solid material that falls out of the air from an erupting volcano. Strictly speaking, it is not at all like the ash that results from a fire. Rather, it is the debris from zillions of bubbles of volcanic glass and fragmented rock. Accordingly, volcanologists have given volcanic 'ash' a completely different name – tephra.

After the Oruanui Eruption, a large lake formed. The lake level was 141 metres above the present lake level, which is about 368 metres above sea level. Remnants of this old lakeshore can be seen in the landscape to the west of the present lake. No doubt the outlet of the lake, feeding the Waikato River, was quickly cut down through the relatively unconsolidated caldera wall, and probably helped by the modifying effects of the 26 subsequent but much smaller eruptions of Taupo prior to the last eruption.

The Next Biggest One: The Most Recent Eruption of Taupo

The Taupo Eruption (note the formal name) occurred about 1800 years ago and produced enough pumice, ash and rock to bury the entire North Island by about a metre, if spread evenly. So its productivity was only 10 percent that of the Oruanui Eruption. Almost as dramatic as the Oruanui Eruption, the Taupo Eruption is the most violent eruption known in the world in the past 5000 years.

Evidence of the Taupo Eruption is abundantly preserved and visible in road cuttings all through the central North Island, especially the Desert Road (Taupo to Taihape) and the Taupo–Napier Road. Forests were destroyed wholesale, resulting in vast quantities of charred logs. If you see charcoal in road cuts in the area, you know you are looking at ash from the Taupo Eruption.

Detailed study has shown that there were seven distinctive stages to the Taupo Eruption, but by far the most dramatic was the sixth stage. About 30 cubic kilometres of pumice, ash and rock were explosively unleashed, producing an eruption column some 55 kilometres high. The column then collapsed to form an extremely fast-moving, incandescent cloud that hugged the ground and raced in all directions across the landscape from the eruption centre, travelling at speeds in excess of 600 kilometres per hour and up to 900 kilometres per hour. Travelling up to 80 kilometres in distance, this pyroclastic flow suddenly stopped and froze to produce a very distinctive rock known as ignimbrite, in this case the Taupo Ignimbrite.

Ignimbrite: A Deadly Cloud

It is hard to comprehend such things! How can volcanoes generate such fast-moving, deadly clouds of pumice, ash and rock? If you imagine a hovercraft racing across flat water on a bed of air, then this is similar to what happens in an ignimbrite.

The material entrained in the eruption column is superheated, in excess of 800°C. The melting point of granite, and hence rhyolite, is about 850°C. Gas-charged liquid rock fountains out of the volcano and rises up through the atmosphere, cooling as it does, forming vast quantities of pumice and ash that have been fragmented while hot (this is what pyroclastic means), and carrying with it rock fragments caught up in the blast from erosion of the throat of the volcano. All this material must then fall back to Earth under the influence of gravity, hence the geological name for these deposits: pyroclastic gravity fall deposits. Simple! What goes up must come down. But what about those tremendous speeds?

A vast amount of material falls back vertically down the sides of the eruption column at colossal speeds. However, when it reaches the ground it can only go outwards; it is forced to race away from the eruption centre and the forceful upward blast of fluid pressure forming the eruption column. It really is a jet. It can only head away from the column, but in so doing it traps air, which in turn is superheated and expands, creating an almost frictionless gas cushion, a magic carpet that literally carries the pyroclastic fall deposit away. In this respect, it behaves rather like a hovercraft, moving apparently without friction. The difference is that it is a deadly transport mechanism, the ultimate razor capable of instantaneously destroying everything in its path.

An excellent demonstration that helps to illustrate this air-trapping process, but which operates with cold air, is to tip a bag of flour on the ground. The flour is so fine and dense that it forms an almost impermeable surface through which the air cannot escape, so the trapped air carries the flour outwards at tremendous speed.

It was Patrick Marshall (Otago University) who recognised the significance of the distinctive rock produced by this process. He named it ignimbrite, meaning 'fire cloud'. Though named from New Zealand, such rock is not confined to New Zealand. Ignimbrites are common enough in the geological record. Spectacular examples are known from Santorini in Europe, Yellowstone in the USA, the Altiplano in Bolivia, Patagonia in Argentina, and in many other places.

In New Zealand, ignimbrites are widespread within and beyond the TVZ. Some are incredibly thick and form high cliffs, such as those near Te Kuiti on the road to Taumarunui. Needless to say, geologists have conferred names on the more conspicuous ignimbrites – that is, those that are mappable, that are extensive

11. Some have described it as evidence of a Polynesian civilisation thousands of years old; geologists have a more prosaic explanation for the Kaimanawa Wall. It is considered an outcrop of Rangitaiki ignimbrite from an eruption 330,000 years ago. As the ignimbrite cooled, it formed into regular shapes which at first glance appear to be man-made.

and/or thick. For example, the ignimbrite-forming, cliffed ridge crests near Eight Mile Junction to the south of Te Kuiti, the Ongatiti Ignimbrite, erupted from the Mangakino Caldera about 1.21 million years ago.

Ignimbrites are very much part of the landscape of the central North Island and relate to the onset of rhyolite vulcanism in the TVZ within the past 2 million years. As such, they provide a superb record of environmental change and landscape evolution in the North Island for the past 2 million years.

Ignimbrite Genealogy
Tracking ignimbrites and deposits of volcanic ash or tephra to individual volcanoes or volcanic centres involves detailed mapping and systematic laboratory work. Yet it is a very rewarding exercise, made easy by the fact that an eruption emanates from a point source effectively, and then, as a general rule, it is distributed outwards in an ever-expanding, radiating pattern away from the volcanic centre. Also, as a general rule, the denser and larger fragmental material is deposited nearer to source than the less dense and smaller-sized fragments. The physical characteristics of volcanic eruption 'products' and their tell-tale distribution patterns are all important in this detective work. But wait! There's more.

Just as in baking, each batch of magma has its own unique chemistry and mineralogy, subtle though the signature may be. So chemical and mineral analyses also provide a means of characterising, identifying and fingerprinting the products of individual eruptions.

The 'Kaimanawa Wall'
One of the most voluminous and widespread ignimbrites in the central North Island is the Whakamaru Ignimbrite, which erupted about 330,000 years ago from the Maroa Caldera. This ignimbrite is exceptionally thick, in excess of 100 metres in places, and forms spectacular cliffs with conspicuous, well-developed cooling structures. These manifest themselves as geometrically arranged, vertical and sub-horizontal fracture planes within the rock. This is a completely natural phenomenon. Cool any solid from a hot state and it will try to shrink.

11

12. On 23 September 1995, Ruapehu came dramatically to the attention of the world when it erupted, disrupting life surrounding the volcano. Eruptions continued through into mid-1996, when gas clouds billowed as high as 11,000 m.

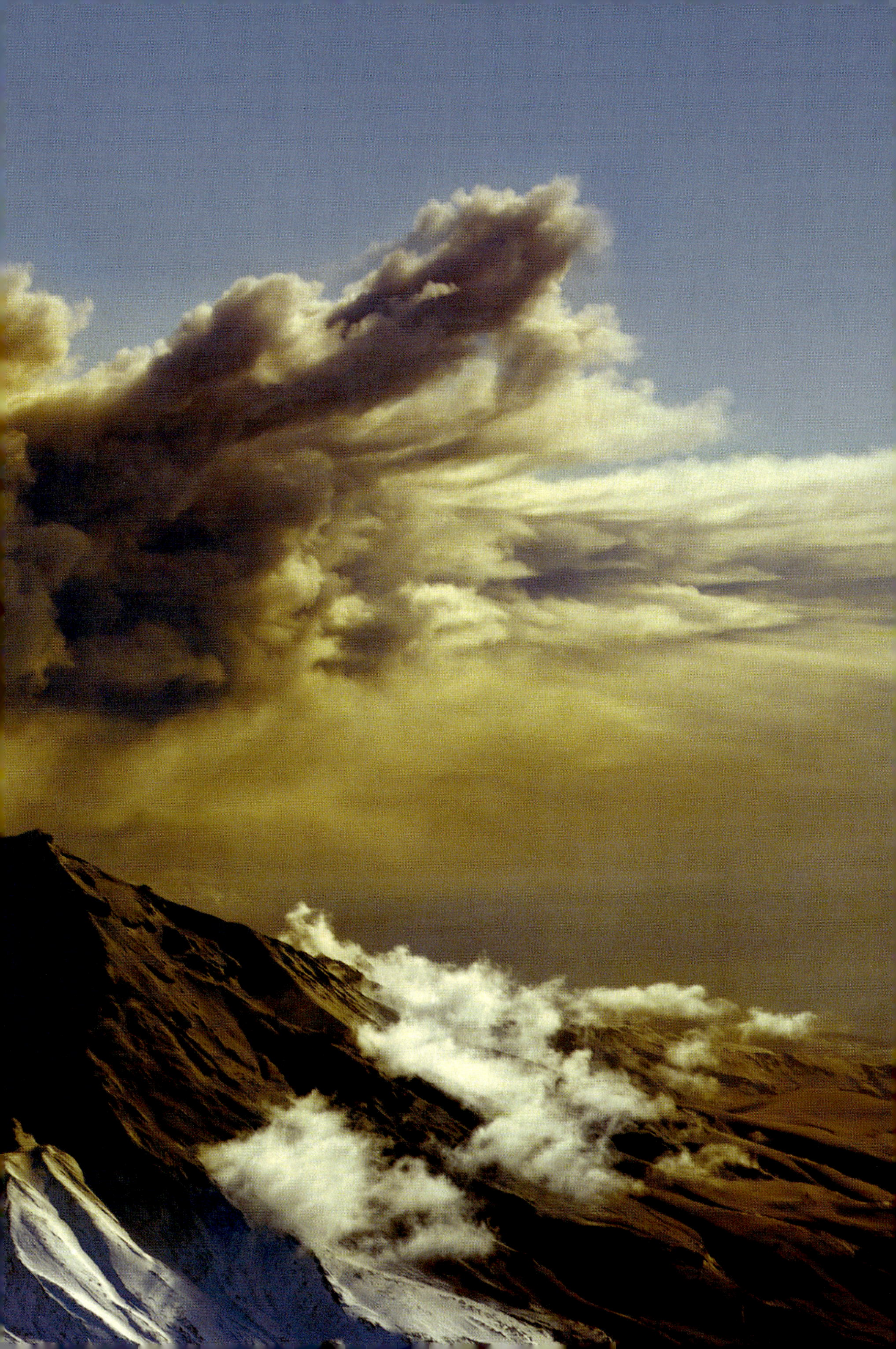

Located on a forestry road about 20 kilometres to the east of Taupo at the northern end of the Kaimanawa Range is the Kaimanawa Wall. It is nothing more than a natural outcrop of Whakamaru Ignimbrite with conspicuous cooling fractures. The fractures are so straight, so clearly defined and so evenly spaced that they look as if they might be man-made. Indeed, they have been compared with man-made stone walls of Inca origin in Peru. However, the Kaimanawa Wall is completely natural.

Several ignimbrites derived from the TVZ have travelled as far as Auckland. It is sobering to think that it would take as little as 20 minutes for an ignimbrite to reach Auckland from Taupo!

Monitoring New Zealand Volcanoes

GNS Science and its monitoring subsidiary GeoNet are keeping track of the Taupo Caldera. In fact, all New Zealand's active volcanoes are being monitored in three ways. First, they are being 'listened' to using seismographs. All seismic activity associated with our active volcanoes is instantly recorded and analysed. Secondly, they are being watched using satellites to see if there is any change in the land surface, any inflation or deflation. This involves geodetic surveying using geographic positioning systems (GPS), as well as other computerised remote-sensing techniques, such as X-ray interferometry using satellite imagery. Thirdly, all major geothermal springs and vents are routinely checked for tell-tale chemical changes that might indicate an impending eruption, such as a marked increase in carbon dioxide, hydrogren sulphide and/or sulphur dioxide emissions. A sudden change in concentration of these volatile gases is often the first indication of something about to happen.

In a sense, the great rhyolite calderas that dominate the TVZ are an expression of intrusion of fresh granite at depth. The vulcanism that has produced the calderas is a near surface consequence. Rifts and granites go hand in hand.

Double Whammy

The situation in the TVZ is made even more interesting because it is also a centre for subduction-related vulcanism, producing andesite. White Island, Tongariro and Ruapehu are the results. In this sense, New Zealand has a double whammy: a volcanic arc superimposed on a rift! There is nowhere else on Earth quite like it. The TVZ calderas or craters are a response to rifting of the continental crust that forms the North Island segment of the Australian Plate. The TVZ cones (such as Ruapehu, Ngauruhoe and Tongariro) are a response to the subduction of the oceanic crust of the Pacific Plate beneath the continental crust of the Australian Plate. It is therefore not exactly surprising that there is so much variety in styles of eruption and magmatic composition all in one small part of the world. Amazingly, the plumbing associated with these various volcanoes is fairly well conserved.

Lava associated with the Taupo eruptions is quite different from that of Tongariro or Ruapehu. It is clear that they are tapping markedly different reservoirs deep within the Earth, and there is no connection between them. Bear in mind that the conduits, pipes or fissures up which magma flows (magma is called lava only once it has reached the surface) are not necessarily very wide. From what we can determine at the surface, the conduit for Ngauruhoe (which at only 2000 years old is the latest cone to have formed on Tongariro), is less than 100 metres in diameter and probably much less.

For all this, there is evidence to suggest that there can be capture of plumbing systems, whereby two markedly different magmas appear to exploit the same conduit. The 1886 Tarawera Eruption is an example of precisely this phenomenon. Whereas the last eruption produced basalt, the previous eruption 700 years before had produced rhyolite. Tarawera Volcano is within the Okataina Caldera that is of similar size to Taupo and has an older history. Lake Rotorua lies within the smaller Rotorua Caldera.

Eruption Triggers

What is it that triggers an eruption? What will make a volcano suddenly blow its top? There is no easy answer, which is why the world needs so many vulcanologists. It all depends on the volcano, and for every active volcano, there ought to be at least one vulcanologist taking a vigilant interest. Each volcano is unique. In the case of the most recent Tarawera Eruption, detailed investigations of what was erupted suggest that the eruption was probably triggered by a basalt magma finding its way into more silica-rich rhyolite magma. Imagine dense, liquid basalt at a temperature in excess of 1100°C interacting with a cooler liquid rhyolite at 850°C. It would be similar to oil and water mixing to produce a highly explosive mix.

The powerful eruption of Krakatoa in Indonesia in 1883 was triggered by basalt magma mixing with cooler rhyolite magma. New Zealand's most active volcano, White Island, is considered to be comparable to Krakatoa. The sudden and violent eruption of Tambora (on Sumbawa Island, Indonesia) in 1815 is also thought to have been triggered by a basalt magma injection. This was the biggest eruption on Earth in the past 10,000 years and is estimated to have been 150 times larger than the Mt St Helen's eruption.

Another trigger might be a simple earthquake. Think of these highly explosive rhyolite calderas such as Okataina and Taupo as high-pressure systems, rather like balloons waiting to be popped. All they need is an appropriate pin to prick them. This is why some geologists became nervous when the Edgecumbe Earthquake struck in 1987. Imagine what might happen if a well-primed magma chamber was suddenly cracked open and breached.

Now that we have addressed some key aspects of the Earth's rocky crust and how it works, including fundamental processes such as vulcanism, we can now delve back billions of years into the very beginnings of the land that we today refer to as New Zealand.

Part 02:

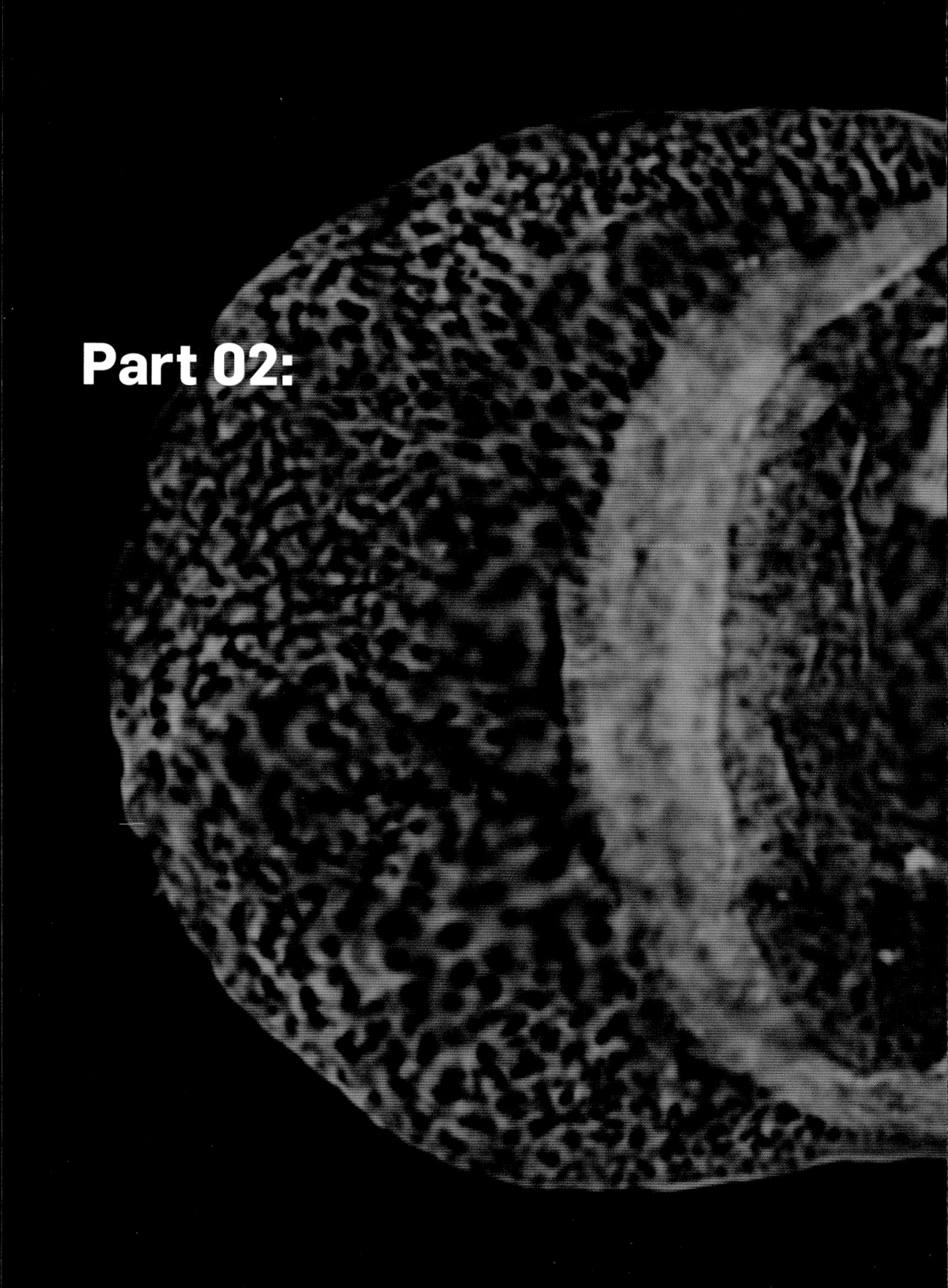

Gondwanaland:
505–83 MILLION YEARS AGO

04/ From Ur With Love: Our Oldest Stuff

From Little Things, Big Things Grow

Although mainland New Zealand's oldest rocks are dated at about 505 million years old, and even older rocks (540 million years old) are found on Campbell Island, the origins of New Zealand can be traced back to the ultimate origin of the Earth itself, 4.53 billion years ago.

Traditional Maori explanations of the origin of Earth appeal to the supernatural, the gods Papatuanuku (Earth Mother) and Ranginui (Sky Father). They are in such a tight embrace that their godly offspring, atua, are in total darkness. The atua pushed their parents apart, creating space and light, and eventually life on Earth became established as we know it today. This is as good an explanation as any other that human societies have developed, if not better. It has strong appeal because it is creative, positive and anthropomorphic. It relates to the central relationship common to all humanity and its continuity, the bond between man and woman, not to mention competition between generations.

By contrast, in the 1600s Bishop Ussher deduced from biblical studies that the Earth was created by God in October 4004 BC. Subsequent ecclesiastical scholars have refined his calculations,

From Ur With Love: Our Oldest Stuff

1. The oldest rocks in mainland New Zealand are of Cambrian age, about 510 million years old, and they are only known from north-west Nelson in the South Island.

2. An aerial glimpse of some of New Zealand's oldest rocks in the Anatoki Range in the central Tasman Mountains of north-west Nelson. These are spectacularly folded strata of Middle Cambrian age.

3

4

and some years later the Vice Chancellor of Cambridge University determined that it happened at precisely 9 a.m. on Saturday, 17 September 4004 BC.

These differing accounts serve two purposes common to all cultures. They acknowledge the certainties that the Earth must have had a beginning, and that there must be an explanation of how it formed. Whether the explanation is right or wrong is not especially important.

Today, it is widely accepted that the Earth is about 4.53 billion years old. This figure has been determined from radiometric dating and is the age of our Solar System. In other words, the Earth is exactly the same age as the Sun and all the other planets: Mercury, Venus, Mars, Jupiter, Saturn, Uranus and Neptune (sorry Pluto!). All available evidence suggests that our Solar System formed as a single entity. Despite searching, no material older than 4.53 billion years has been found.

The oldest objects known to us are meteorites. These are pieces of rock carrying messages from the past that fly within our Solar System and scorch through Earth's atmosphere. By dating meteorites we can determine the age of our Solar System.

The Auckland Meteorite

At 9.30 a.m. on Saturday, 12 June 2004, a meteorite crashed through the roof of Brenda and Phil Archer's suburban house in Ellerslie, Auckland. Mrs Archer was making porridge in the kitchen, just a few metres away from the living area where the meteorite came through the ceiling. Alarmed by the explosion, she yelled to her husband. They soon realised that the noisy intruder was a meteorite about the size of a cobble. It was smooth, grey-brown in colour, rather like a large discoidal potato, and was still warm. It weighs 1.3 kilograms. There is a small chip where it presumably impacted with the metal roof. It punched its way straight through the roof and ceiling, and landed on the living-room sofa, which behaved like a trampoline. The meteorite bounced up, bruising the ceiling, and then clattered to the floor. Only minutes earlier, their grandson had been playing in the room.

Now at the Auckland Museum, the Auckland meteorite is one of only nine meteorites known from New Zealand. It appears to be an ordinary chondrite (L6 or LL6 in meteorite jargon). It has not been analysed or dated, but it is interesting to speculate that it may well be as old as the Earth.

How do we know this? And how can we be so certain? It is largely to do with our ability to determine the age of rocks, and it all started with Ernest Rutherford. It was he, along with his colleague Frederick Soddy, who in 1903 suggested the theoretical possibility of dating minerals, and hence rock, using radioactive elements such as uranium and potassium.

New Zealand's Mysterious Zircons

The oldest dated minerals known from New Zealand are zircons that are in excess of 3 billion years old. They predate the first continent, Ur, which dates from about 3 billion years ago. These zircons were dated by Chris Adams using a mass spectrometer at the GEMOC Laboratory at Macquarie University,

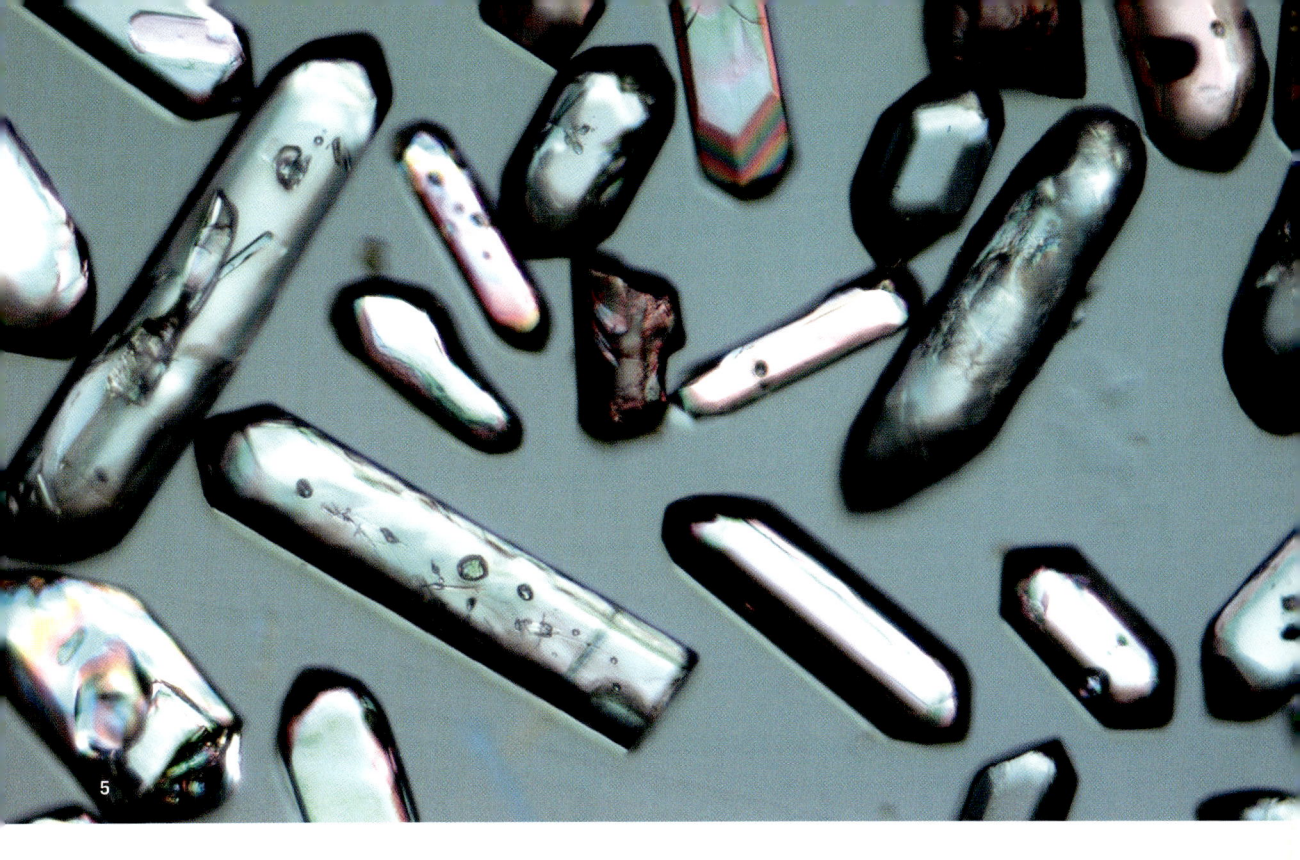

3. This meteorite was a surprise intruder into an Auckland house on 12 June 2004. Quite possibly as old as our Solar System at 4.53 billion years, it is one of only two meteorites recorded to have fallen in New Zealand. At least seven other meteorites are known from New Zealand, but these have been lucky finds.

4. Lying 53 km to the north-west of Cape Reinga, the Three Kings islands are not only home to rare plant species such as *Tecomanthe speciosa*, but also New Zealand's oldest-known minerals. Zircons extracted from greywacke rocks on the Three Kings are more than 3 billion years old.

5. Zircon is a mineral that is easily dated because it contains uranium. These are typical crystals of zircon, less than 5 mm long that form as a common but minor mineral within granite.

Sydney. The zircons were extracted from a sample of greywacke collected from the Three Kings Islands to the north-west of the North Island of New Zealand. This rock was collected by two geologists from the University of Auckland, Philippa Black and Bernhard Spörli. No doubt older zircons will turn up from elsewhere in New Zealand, but for now these are the oldest.

What the zircons' ages tell us is that at least some of the original sediment that makes up the greywacke on the Three Kings Islands is derived from extremely ancient rocks. The zircons do not tell us where they came from; all they reveal is how old the rock is that they were derived from. However, there are no zircon-bearing rocks in New Zealand or eastern Australia from which such old zircons could be derived. The mystery deepens! The nearest known source is in northern China. Could this be the source of these old zircons? In the absence of a better answer to this riddle, it is indeed a reasonable suggestion. There is no basis for excluding this possibility.

Einstein Explains How Radioactivity Works
In 1905, just two years after Rutherford and Soddy had pronounced how to date rocks, Albert Einstein explained how radioactivity works. In so doing, he established the relationship between mass and energy, invoking his immortal equation $E=mc^2$. He wondered about the source of the radiation energy emanating from a radioactive element. He realised that as a result of radioactive decay there was a loss of mass, yet the radiation energy (alpha, beta and gamma particles) has virtually no mass. It slowly dawned on him that the amount of energy involved in a single atom is colossal. Radioactivity is the spontaneous transformation of mass into energy. Only some elements do this, the most common of which are uranium and potassium.

Einstein also established that the slower the rate of decay, the less energy is produced; whereas the faster the decay, the larger the energy produced. The energy is released by reducing the mass, either by fission or by fusion. In so doing, he explained how the Sun works, and this has to be one of the greatest breakthroughs in human comprehension. If you take four hydrogen atoms and squeeze them (by fusion) to

6

7

6. Albert Einstein and physics colleagues at the first Solvay Conference held in Brussels, Belgium, in 1911. These are some of the scientists who sorted out our modern understanding of the nature of matter, the age of the universe and its origins. From left to right, standing: Goldschmit, Planck, Rubens, Sommerfeld, Lindemann, de Broglie, Knudsen, Hasenorhl, Hostelet, Herzen, Jeans, Rutherford, Poincaré, Einstein, Langevin. Sitting: Nernst, Brillouin, Solvay, Lorentz, Warburg, Perrin, Wien, Curie, Kamerling Onnes.

7. Reading the rocks: Chris Adams (GNS Science) is New Zealand's leading authority on the dating of minerals. He is sitting in front of a mass spectrometer devoted to analysing isotopes of argon for dating potassium-bearing minerals.

form one helium atom, what you get is an overall reduction in mass and the release (radiation) of a vast amount of atomic energy: sunlight.

Reading the Rocks

An Englishman from Bedfordshire, and a graduate of London and Oxford universities, Dr Chris Adams came to New Zealand in 1969 and quickly built New Zealand's first potassium-argon mass spectrometer. Today he is New Zealand's best-known geochronologist and is responsible for dating thousands of New Zealand rocks. After two decades of work dating igneous and metamorphic rocks, he has latterly turned his attention to the dating of detrital minerals (sand grains) in ancient sedimentary rocks. The age of detrital minerals gives the age of the parent rock from which they are derived. This kind of information is extremely useful in determining the provenance or source of the original sediment.

Zirconium silicate, or zircon, is a common but minor mineral in granite. It happens to also be the most significant uranium-bearing mineral in granite, but there are others such as monazite and apatite. As the name suggests, the main element in zircon is zirconium, but a variety of other elements with similar mass can occupy the same crystal lattice sites as zirconium within the zircon crystal, and in particular uranium.

Zircon is a remarkably robust mineral and is known and prized for its toughness and durability. It is usually translucent or transparent and can form large crystal masses. It is regarded as a semi-precious stone and is widely used for jewellery. However, for scientific purposes, the zircons extracted from rock samples are very small, less than 3 millimetres in length, and usually less than 1 millimetre. Think of them as sand-sized grains.

How to Capture Zircons

First, the rock sample is selected. Fresh rock is preferable to weathered rock – about 300–500 grams will do. It is then reduced into gravel-sized fragments ready to be crushed, using a TEMA tungsten steel-crushing mill. It takes less than a minute to crush rock in a machine like this. It behaves rather like an orbital sanding machine, except what is being moved about is a squat, round, steel saucepan-shaped vessel with a secure lid and at least one free-moving steel ring or disc inside. The relative motion between this ring or disc and the wall of the vessel is what crushes the rock fragments.

The rock sample is crushed to form sand. If the rock is crushed for too long, it will be crushed to silt or even finer grain sizes, which is no good for this purpose. What is required are mineral grains 1–3 millimetres in length. Sand-sized particles are ideal.

The crushed material is then washed through sieves to remove the unwanted large and small particle sizes. Next, the sand is dried and passed through a strong magnet. This will remove a significant portion of unwanted sand, all the iron-bearing minerals. Then the residual sand is tipped into a beaker of heavy liquid solution of sodium polytungstate. All the light, non-magnetic, low-density minerals such as quartz and feldspar will float and can easily be disposed of. The remaining residue contains heavy, non-magnetic, higher-density minerals, including zircon. This is rinsed, dried and then the zircons are hand-picked with the aid of a binocular microscope.

The zircons are then individually mounted on a glass slide, ready for placement in a vacuum chamber within a mass spectrometer. The zircons are now ready to be zapped with a laser beam. The beam is about 5 microns across and burns a neat pit into the zircon crystal, instantly breaking the chemical bonds to create an ionised plasma, or stream of free-flowing atoms. This stream of atoms is drawn off and passed through the mass spectrometer for immediate analysis. The instrument is calibrated to detect and count uranium and lead atoms. The machine is so sensitive that it can detect all the various isotopes of uranium and lead, of which there are many. Armed with all this clinical data, an on-board computer then automatically calculates the age of the zircon. The process is all very fast. A single zircon takes about 10 minutes to process within the mass spectrometer and compute a date.

05/ The Greywacke Story

Greywacke And Schist

Every New Zealander should know greywacke. It is the most common rock in New Zealand, forming about 60 percent of the New Zealand land mass. The word is derived from an old German mining term for rather nondescript grey sandstone or 'grauwacke'. This was the boring, uninteresting, unmineralised and unwanted rock material that miners had to laboriously excavate in order to reach the good stuff, the ore. Greywacke is a sedimentary rock best described as 'dirty sandstone'. And, yes, it is usually grey! It started off as sediment accumulating on the sea floor, dominated by sand with a proportion of silt and mud.

However, it is more than a sedimentary rock, because it is also metamorphosed. It has been buried and slow-baked at elevated pressures. This process has turned the original sandstone into greywacke. With increasing pressure and temperature, greywacke will become schist. Predictably enough, much of the schist in New Zealand is metamorphosed greywacke.

Greywacke and schist form the axial ranges of the North and South islands. It is especially conspicuous in the South Island, forming much of Marlborough, Canterbury and Otago, and the Southern Alps.

The Greywacke Story 65

1. Greywacke dominates the geological makeup of New Zealand. Geologists call it the 'basement' rock of New Zealand. It is our most widespread and most characteristic rock. Schist is a metamorphosed (more deeply buried and cooked) form of greywacke.

2. A world dominated by greywacke. In this view of inland Canterbury, the snow-capped hills of the Mt Hutt Range (to the left) and more distant ranges to the north (to the right) of the Rakaia River (foreground) are the source of the extensive greywacke gravels that form the plains.

3. The Kaimanawa Range, an axial range within central North Island, looking west with the southern slopes of Ruapehu just visible in the distance at top right.

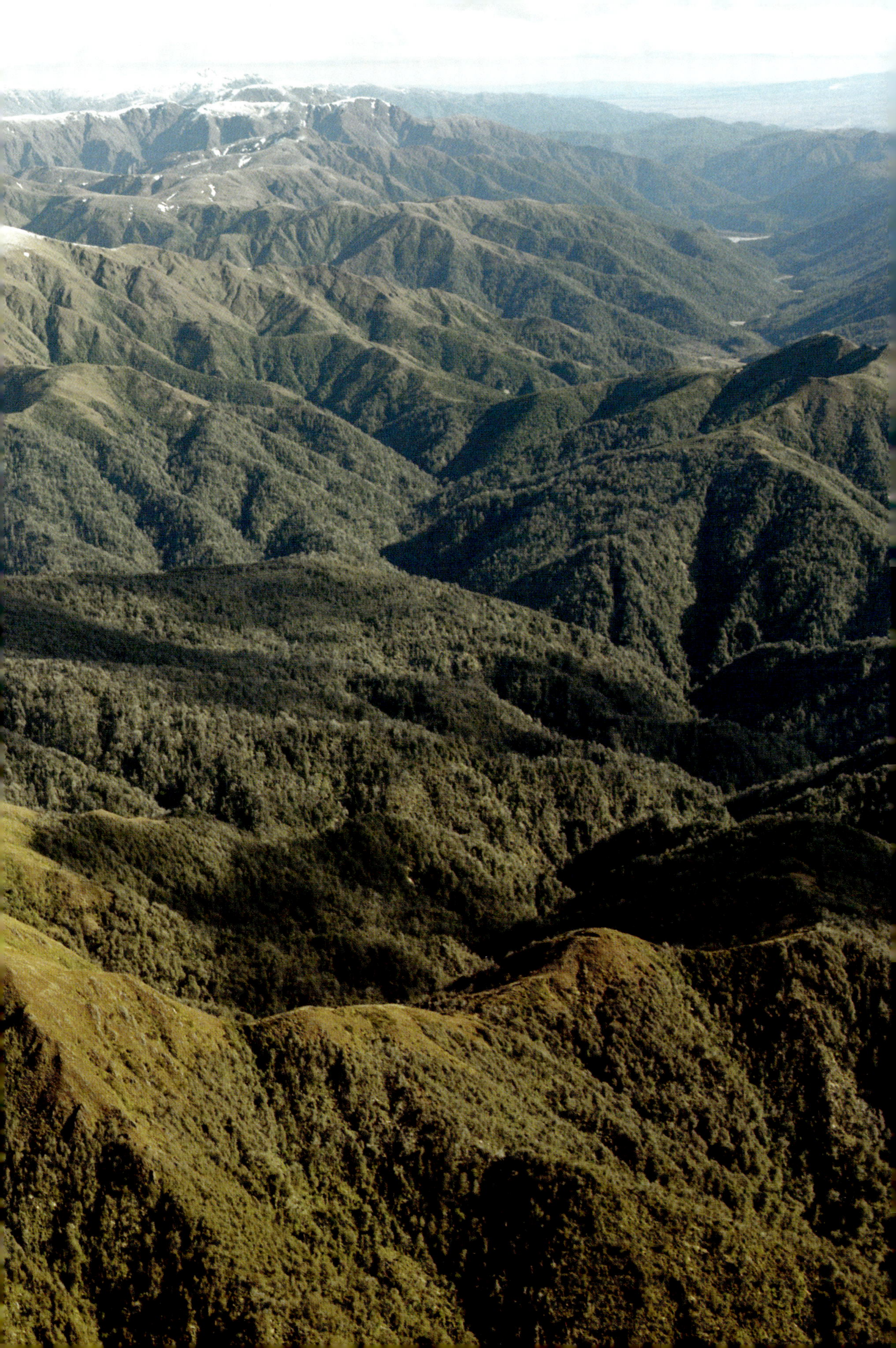

Schist is widespread in Otago, the west side of the Southern Alps and Marlborough. It is much less widespread in the North Island, but is known in a few areas such as Terawhiti (west of Wellington) and the Kaimanawa Range (east of Ruapehu). Schist also occurs in the Chatham Islands and Campbell Island.

Not only do greywacke and schist form the principal basement rock of New Zealand, they are also the main source of younger sedimentary rocks and sediments that form so much of our landscape. This includes the so-called 'papa' of the North Island and the voluminous gravels of the Canterbury Plains in the South Island. Most of our major rivers are transporting and depositing reworked greywacke and schist. Inevitably, much of the sand and gravel forming the riverbeds and beaches of New Zealand are derived from greywacke.

There are, of course, exceptions, such as where there is granite, volcanic rock or limestone. However, greywacke is the rock that dominates our landscape.

Source of the Greywacke

Greywacke is a sedimentary rock, which means it started off as sediment, which in turn is derived from pre-existing rock. Two key questions arise: where is the source of the original greywacke sediment; and what was the pre-existing land that gave rise to the sediment?

You might think the answer to the first question is obvious and that the source is nearby, somewhere within the New Zealand region. But it is not obvious at all. For decades, New Zealand geologists have been searching for a suitable answer to this question, and we do have some very strong leads.

First, we know what the parent rock of the greywacke is or was. From exhaustive studies of the mineral composition of the greywacke, we know that the major source was granite. The original detrital sand grains that make up the greywacke are exactly what you would expect if you ground up granite and dumped it in the sea: a mixture of quartz, feldspar and mica, with various other common but much less abundant minerals such as zircon.

4. The Southern Alps: fashioned from greywacke. Plate collision is responsible for their uplift along the eastern side of the Alpine Fault. They have formed along the western edge of the Pacific Plate. Immediately adjacent to the Alpine Fault, immense pressure has transformed the greywacke into schist. Modern surveying using satellites and lasers shows that the Southern Alps are rising up to 10 mm per year, and are being eroded almost as fast. This view is near the Garden of Eden Ice Plateau.

Although there are granites exposed in the South Island of New Zealand, they prove not to be the source of the greywacke sediments, for several reasons. For starters, most of this granite is 125–105 million years old (Late Cretaceous in age) and hence younger than the greywacke. Clearly it is impossible to have a rock source that is younger than the sediment. The greywacke is 490–140 million years old, of Ordovician, Carboniferous, Permian, Triassic, Jurassic and Early Cretaceous age. There is also older Devonian granite, but this is too small a body to produce the vast amount of sediment represented by the greywacke. Amazingly, there is no suitable source for the original greywacke sediment within the New Zealand land mass. Yet it had to have come from somewhere.

It could be that the source is hidden from us, or is totally eroded away. These are possibilities, but they are considered to be remote and extremely unlikely explanations, because we know too much about the geology of New Zealand, Zealandia, Gondwanaland and the ocean floor.

From all available evidence, we can be certain that the greywacke source rock was located within continental crust of eastern Gondwanaland. Today, the source area or areas must be located in Antarctica and/or Australia. This makes sense in that Zealandia broke away from eastern Gondwanaland prior to separation of Australia from Antarctica. Both Australia and Antarctica were major sectors of eastern Gondwanaland.

New Zealand's Greywacke: A Queensland Legacy?
In order to determine the nature of the mother rock that the greywacke sediments have been eroded from, hundreds of greywacke samples have been carefully selected from all over New Zealand. What we mean by 'carefully selected' is that they have been chosen in such a way that we can be sure that they are representative. They are not in any way special. Rather, they are typical. They are all fresh; that is, as unweathered as possible – the minerals are not rotten.

And they are unaltered; that is, they are free of mineral veins (especially quartz and zeolite veins) and there is nothing to suggest that they have been subjected to hot fluids passing through them. Furthermore, the samples selected are all of similar grain-size: they are all medium sandstones.

The reason for being so selective is that this is a comparative study, and we need to ensure that we are comparing like things. By comparing samples of more or less identical attributes from different greywacke throughout New Zealand, we can be certain that we are comparing apples with apples, and not apples with oranges.

Mother Nature is a great natural sorting machine. However, some caution is required. A medium-grained sand, as opposed to a fine-grained or coarse-grained sand, will tend to have slightly different concentrations of some minerals compared with a siltstone, for example. This is because sediment grain-size depends on the physical and chemical attributes of the constituent minerals involved: minerals vary in crystal structure, shape, size and strength or hardness. When selecting a sample for analysis of the sand grains within it, we tap into and exploit nature's selection process.

For all samples, the zircon and mica minerals are extracted using standard mineral separation techniques. Because minerals vary in chemical composition, they vary in density, and it is relatively easy to separate minerals using fluids of different densities. A centrifuge helps!

A subset of zircon crystals (and/or mica crystals) is then hand-picked, and each one is individually dated. A population of up to 100 grains is dated for each sample, and then the dates are plotted on a graph. This produces a 'snapshot' of the age distribution of the zircons and/or micas in the sample, and tells us at a glance the ages of the zircon-bearing source rocks that the greywacke sediments are derived from. This approach is based on the premise that the age of a zircon crystal is a reliable indicator of the age of crystallisation of its granite host rock.

The age distribution of zircons and micas from New Zealand greywacke consistently suggests that the majority of zircons are of Permian–Triassic age, with subsidiary Ordovician micas (but no Ordovician zircons). What this means is that the source rocks are granites of largely Permian–Triassic age, with subsidiary schist of Ordovician age. The schist is mica-bearing but not zircon-bearing.

The key is to establish the whereabouts in eastern Gondwanaland of granites and schist of the right age to produce this particular detrital mineral age signature. After exhaustive research, Chris Adams and colleagues have determined that there is nowhere suitable in New Zealand, Antarctica, Tasmania, south Australia or south-east Australia. However, there is a possible source in north-east Queensland. This hypothesis was first expressed in the mid-1990s, and since then considerable efforts have been made to test it. However, all new work has confirmed this interpretation. It is stronger than ever.

It is funny to think of the bulk of the basement rock of New Zealand, the greywacke, perhaps our most distinctive rock, as effectively a large piece of discarded Queensland real estate! As Chris Adams says, 'The Australian heritage of New Zealand includes Vegemite and an awful lot of assorted muddy sandstones from Queensland.'

In New Zealand, there is an enormous range in age of the greywacke, stretching from the Ordovician to the Early Cretaceous, a span of more than 350 million years. Not surprisingly, there is greywacke and greywacke. Geologists have found ways of distinguishing one from another, using a variety of characteristics, age being just one of them.

Age of the Greywacke

When geologists talk about the 'age of the greywacke', what is meant is the age of deposition of the original sediment on the sea floor. This age is determined from fossils. The process might sound easy enough, but fossils have to be found first, and then they have to be identified so that the age can be determined. Both of these steps require expertise. Geologists are trained to recognise and collect fossils, but

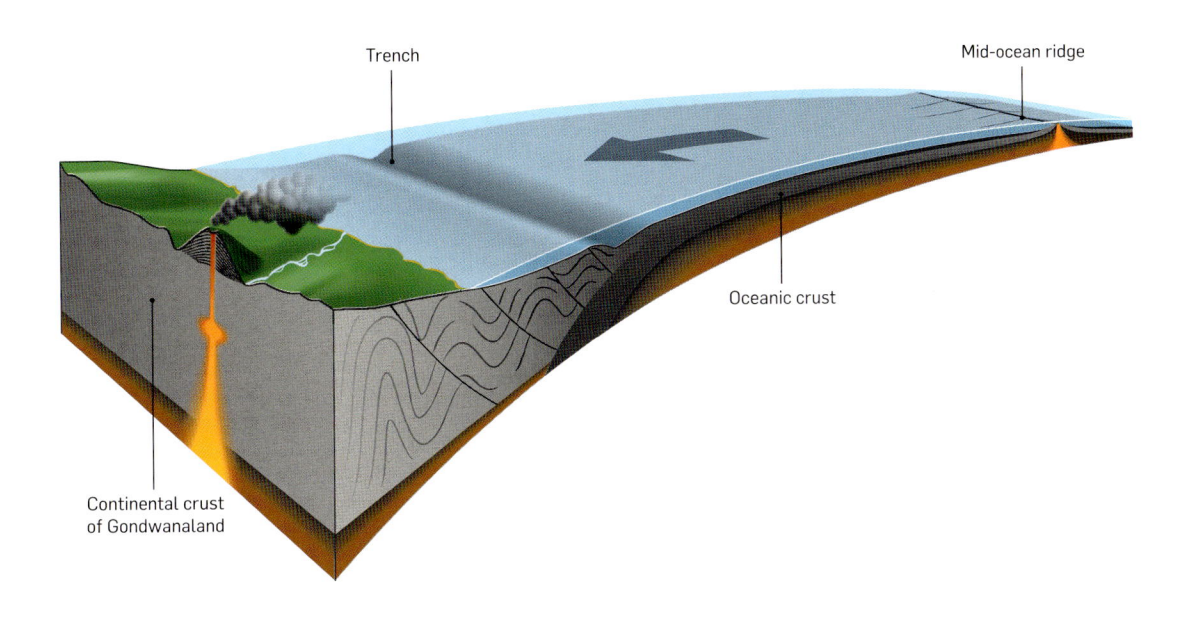

Labels on diagram: Trench, Mid-ocean ridge, Oceanic crust, Continental crust of Gondwanaland

5. Continental growth by accretion. The original greywacke sediments were eroded from granitic rocks on Gondwanaland, transported by rivers, and deposited on the sea floor. As the sediments arrived, they were bulldozed by sea-floor spreading and scraped off, piled up and accreted to the eastern margin of Gondwanaland. This same process is occurring today around the active subducting margins of the Pacific Plate, such as the eastern North Island.

not all are equally skilled at doing this. Some are better than others. Paleontologists are specialist geologists who are trained in the identification of fossils and the determination of their age.

The Oldest and Youngest Greywacke

Our oldest greywacke is referred to as the Greenland Group and is of Ordovician age, 490–443 million years old. It is widespread on the West Coast of the South Island and is best seen along the coast north of Greymouth towards Westport. Appropriately, but quite coincidentally, the rock is green not grey. The age is based on graptolite fossils from only one locality in the Waitahu River near Reefton. But one single fossil locality is enough.

Our youngest greywacke is of Early Cretaceous age (around 140 million years old) and is widespread in eastern New Zealand, especially Marlborough and the east of the North Island. Numerous shelly fossils are known from these rocks, along with plant fossils, microfossils and vertebrate fossils. By far the most common fossils are shells and shell fragments of a distinctive, large clam named *Retroceramus*. The next most common fossils are shells of another group of bivalves: the trigoniids.

Greywacke Forms the Continent Zealandia

It is worth noting that New Zealand's oldest greywacke occurs along the western edge of our land mass and the youngest greywacke is on the eastern edge. This arrangement makes perfect sense in terms of the natural and predictable history of the growth of a continent. In this case, the greywacke rocks of New Zealand relate to Gondwanaland. They predate Zealandia and, of course, New Zealand. They are part of our inheritance from Gondwanaland.

With the long passage of time, sediments eroded off the continental crust of eastern Gondwanaland. They accumulated in

6

7

6. Gondwanaland as it may have looked 200 million years ago at the very end of Triassic time. This huge continental configuration included future Australia, Zealandia, Antarctica, India Africa and South America. Its eastern margin faced the Panthalassa Ocean.

7. The most common fossil in greywacke of Late Triassic age is a tubular fossil named *Torlessia*. Named after the Torlesse Range in Canterbury, this fossil is considered to be a relative of modern-day single-celled, tube-forming organisms called *Bathysiphon* (foraminifera) that live on the sea floor in water depths of more than 200 m. From this we can say that much of the original greywacke sediments accumulated in water depths in excess of 200 m.

the ocean along elongate basins parallel to the eastern margin of the continent, with the older sediments to the west and progressively younger sediments to the east. These elongate basins formed parallel to the eastern Gondwanaland margin as a consequence of sea-floor spreading and subduction. Oceanic crust of the Panthalassa Ocean was moving westwards (and eastwards), just as part of the Pacific sea floor is today, and was being sucked down (subducted) under the leading eastern edge of Gondwanaland. During this process, elongate basins formed as a series of long, sub-parallel wrinkles between the trench and the edge of the continent.

Sediments derived from the continent were flushed out to sea by rivers, only to become entrained in a vast submarine shunting operation, redistributing sediment into various basins. Some basins received sediment simultaneously; others did not. Some were filled faster than others, and then the sediment spilled over into adjacent basins.

From detailed study it has been possible to determine those that were shallow and those that were deeper, those near land and those more remote and oceanic. The history of each basin is unique, and yet they are all inter-related for one very good reason: they shared the same continent and the same ocean.

With the relentless eastward transport of sediment off Gondwanaland into north–south oriented basins, coupled with the relentless westward motion of the Panthalassa Ocean floor, the intervening basins and their sediments were bulldozed and added to the eastern margin of Gondwanaland in a remarkably orderly fashion. This process is referred to as 'continental accretion' or 'continental growth'. The greywacke in New Zealand is a record of the slow marginal growth of continental Gondwanaland from Ordovician Period to Early Cretaceous time, spanning 350 million years.

Triassic Greywacke

We have considered the oldest and the youngest greywacke in New Zealand, but by far the bulk of it is of Triassic age, 251–199 million years old. On the basis of fossils, we can be certain that there is also Carboniferous, Permian and Jurassic greywacke. However, in terms of volume, the Triassic takes the prize. There is masses of it, and most of it is Late Triassic, 235–200 million years in age.

This poses an obvious question: what was it about the Triassic that promoted such voluminous production and deposition of sediment? There is no easy answer. By far the most common fossil in the Triassic greywacke is a rather simple 5–10-centimetre-long tube that curves and tapers slightly. It is named *Torlessia* after the formal geological name of much of the greywacke: the Torlesse Supergroup. This in turn takes its name from the Torlesse Range of mid-Canterbury, which is best seen from the road to Arthur's Pass.

Torlessia is referred to by geologists as a 'tube fossil' and is interesting because the tube was not secreted by the animal as a structural entity like a shell. Rather, the tube is constructed of grains of sand that the organism appears to have selected and stuck together using powerful organic glue. There are modern-day organisms that live on the deep ocean floor that produce tubes just like this. These are called *Bathysiphon*. *Torlessia* is just one example of a huge variety of single-celled animals called foraminifera, a large group of diverse and extremely abundant marine organisms that are very near the base of the food chain. Foraminifera that build tubes from sand particles are referred to as 'agglutinating foraminifera'.

If you know what you are looking for, tube fossils are surprisingly common in greywacke of the Canterbury and Wellington areas.

06/ The Oldest Rocks: Western Province

New Zealand's Basement Rocks: The Foundation
When the first New Zealand rocks we know of were created more than 500 million years ago, it was a topsy-turvy world compared with today. At that time the supercontinent Gondwanaland – made up of today's South America, Africa, Australia, India, Antarctica and Papua New Guinea – was in the northern hemisphere, not the southern. New Zealand was then about the same distance from the North Pole as it is today from the South Pole – around 5000 kilometres.

It was a period of enormous change on Earth, marked by an explosion of life forms. The first organisms had been simple bacteria and algae that developed 3.5–2.5 billion years ago. By 650 million years ago, soft-bodied animals such as jellyfish, worms and sponges had evolved, but it was only in the Cambrian Period (starting 544 million years ago) that animals with skeletons and shells first appeared. This chapter examines what we know of our oldest Gondwanaland inheritance.

Our Gondwanaland history embraces all New Zealand rocks and fossils that predate the separation of Zealandia 83 million years ago. The time span involved is a colossal 420 million years, stretching from 505 million years ago to 83 million years ago, from Middle Cambrian to Late Cretaceous time.

New Zealand's oldest rocks are referred to as 'basement' rocks – first laid down over 500 million years ago – and occur in the west of the South Island. They are known as the Western Province, while the younger basement rocks of the eastern South Island and the North Island are known as the Eastern Province.

Striped Terranes: Continental Growth Rings

One of the most striking aspects of New Zealand geology is best seen on a map with everything less than 83 million years old stripped off, along with some of the obvious fault deformation removed. In other words, only the old rocks are shown. It shows the foundations of modern-day New Zealand, our basement. What is immediately apparent is that there is a succession of parallel belts of rock. Each one is referred to as a 'terrane'. This is geological jargon, differing in spelling from the geographic term 'terrain'. A 'terrane' in geology refers to a fault-bounded body of rock with a unique history. So a terrane has a different history from adjacent terranes.

This concept was introduced in the 1970s and has proved extremely useful. It has greatly helped us understand how continents grow at their edges. They do so by incremental accretion along their margins. In New Zealand, each terrane has been sequentially added from west to east. So the oldest terranes are to the west and the youngest terranes to the east.

In this respect, Zealandia, and hence New Zealand, are manifestations of 420 million years of deposits along a segment of the Gondwanaland margin. The linear parallel terranes on the map resemble growth lines on a tree or a shell, and this is exactly what they are: incremental lines of growth.

New Zealand terranes vary in dimension, but are mostly narrow, from a few kilometres to tens of kilometres wide. However, they are very long: hundreds to possibly thousands of kilometres long. Clearly these elongate strips of crustal rocks have formed up against a long shape – and the only candidate is the margin of a large continent.

From all our acquired knowledge of how the Earth's crust behaves, this is exactly what we might expect. Zealandia, and hence New Zealand, is a long thin slice carved off the edge of Gondwanaland.

On closer inspection, there is no evidence of any rock formation in New Zealand that came from the actual continental Precambrian craton (the early, stable part of the Earth's crust) of Gondwanaland. So although we may be a 'chip off the old block', the entire chip is from the accreted edge of Gondwanaland! Having said this, we do have sediments that are undoubtedly derived from the Precambrian craton of Gondwanaland.

Terrane Identity Using Strontium

Each terrane has its own unique history in time and space, and inevitably their individual histories are complex. After all, they have been around for a long time and they have all been subject to significant burial, metamorphism, deformation, uplift and erosion. We know about them by virtue of the fact that they are now exposed at the surface. However, their detailed histories are all too complicated to unravel in this book. What is important here is to understand how we know about their histories. From careful analysis of the rocks and the fossils they contain, we can determine how old they are, where they came from, and by what processes.

In the past few decades, we have discovered a successful way of characterising terranes by looking at their strontium content. Strontium is a minor but nevertheless common radiogenic element that is similar in size to calcium, so it invariably occurs in calcium-bearing minerals. When rock is buried and subjected to pressure-cooking at elevated pressures and temperatures, the chemical ingredients homogenise and reorganise themselves. Some do so more readily than others, and strontium is one of them. It is as if all the strontium atoms in the rocks start linking arms, so to speak, and start communicating with each other, redistributing themselves evenly through the rock.

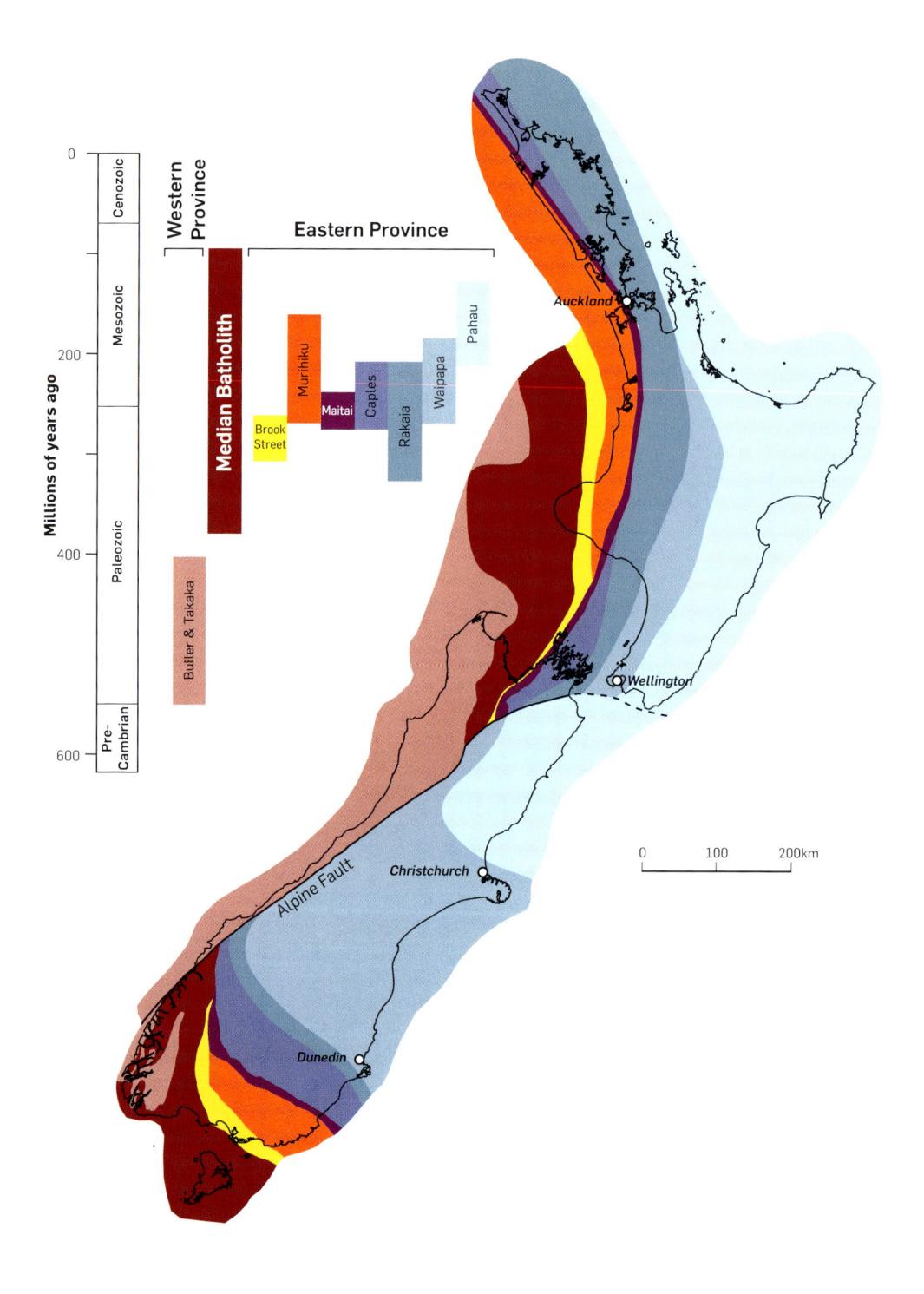

1. Paleozoic–Mesozoic rocks form the basement or solid rock substrate of New Zealand. They can be described and mapped in terms of two provinces, lying to the east and west of a belt of igneous rocks (mainly granite) referred to as the Median Batholith. Each province comprises a series of terranes, elongate belts of rock, each with its own unique history and separated one from the next by a major fault.

2. Trilobites were a large type of marine slater, so named because of a three-lobed body. These armoured arthropods first appeared around 540 million years ago and became extinct by 250 million years ago – an extremely long time for a class of animals to survive. Their tough exo-skeletons remain as fossil proof of their existence and provide vital clues to geologists researching Earth's history. This limestone is stuffed full of skeletal trilobite remains, mainly of a type called *Ptychagnostus*.

A suitable analogy would be baking a cake. Among the ingredients may be half a teaspoon of salt or a dash of vanilla essence. Now, cook the cake and the end result has a uniform taste and texture. Those tiny quantities of salt and vanilla essence have been evenly distributed throughout the cake during the cooking process. The same happens for strontium in rocks. We can now think of each terrane as a unique 'cake' with its own distinctive 'taste'. By determining what is referred to as 'the initial strontium ratio at the time of metamorphism', we have been able to fingerprint all of our terranes.

The name of this technique sounds complicated, but that is for a very good reason. Strontium is radiogenic, so it changes (decays) with time. Therefore, it is necessary to specify a moment in time that is most suitable for comparisons between terranes. An easily determined moment in the history of a terrane is the age of metamorphism.

The age of metamorphism is easily determined by dating new minerals that grew or crystallised when the rock was metamorphosed. Using a microscope, it is usually easy to distinguish between the original minerals in a rock and the new ones. The scientists who have mastered this strontium technique are Chris Adams and Ian Graham, both with GNS Science, and it works brilliantly. We can discriminate between terranes just by interrogating the strontium.

To Source Pounamu

Pounamu, or New Zealand jade, is a metamorphic rock. Pounamu includes two particular types of rock: the more common nephrite; and less common (bluer and softer) bowenite or tangiwai. As luck would have it, the known *in situ* sources of these comparatively rare rocks fall within three different terranes and they all have different strontium isotope signatures.

By analysing a piece of pounamu for its strontium isotope signature, it is possible to fingerprint it and determine its source. This will be a very useful tool for verifying authenticity. Using this technique, it is not only easy to discriminate between New Zealand jade and jade from other parts of the world, such as Australia, China and Canada, but also to discriminate between known occurrences in New Zealand.

The Provinces: Groupings of Terranes

Looking at the map, the basement terranes of New Zealand fall neatly into two groups: the Western Province; and the Eastern Province. They are separated by a belt of rock referred to as the Median Batholith.

The Western Province includes two terranes named Takaka and Buller. These two terranes include the oldest rocks in mainland New Zealand, the oldest rocks that we have inherited from Gondwanaland.

The oldest rocks we know of from Zealandia are actually on Campbell Island, the island group part way to Antarctica. These are of Precambrian age, greater than 542 million years old, but not by much. These rocks are referred to as the Complex Point Schist and they occur more or less at sea level, forming the shore platform in Garden Cove. We have determined the age of metamorphism of this schist, but we have also determined the age of the original sediments using uranium-lead dating of zircons.

Western Province: Most Like Tasmania

The Takaka and Buller terranes include rocks of Cambrian age (542–490 million years ago), Ordovician age (490–443 million years ago), Silurian age (443–417 million years ago) and Devonian age (417–359 million years ago), and they are capped by tell-tale remnants of cover rocks of Permian age (299–251 million years ago), Triassic age (251–199 million years ago) and Jurassic age (199–145 million years ago) that all show a strong resemblance to rocks of the same age in Tasmania. They are quite different from the more widespread Permian, Triassic and Jurassic rocks and fossils of the Eastern Province.

Late Permian rocks of the Parapara Group, with fossil shellbeds rich in brachiopods, bryozoans, bivalves and gastropods, are exposed high on Parapara

Peak overlooking Golden Bay. Triassic sediments of the Topfer Formation with Middle to Late Triassic palynomorphs are found in dense bush in the Waitahu River area near Reefton. Associated with the Triassic sediments is a distinctive Jurassic dolerite that is remarkably like the Ferrar Dolerite, a famous large igneous province (LIP) or flood basalt that is widespread in Tasmania and Antarctica. These three occurrences are indeed very restricted in extent, yet they speak volumes about the history of the Western Province.

Because of these stronger links with rocks of similar age in Australia (particularly Tasmania), the Western Province is sometimes regarded as part of the 'foreland' of Gondwanaland. This is not an unreasonable assumption. The younger Eastern Province terranes are generally considered as being a little more remote and distal to the continental edge of Gondwanaland, and so they should be. This is exactly what we might expect in an orderly progression of continental growth seawards of a north–south-orientated continental margin with the open ocean to the east.

Let us now consider the fossil record of the Cambrian to Devonian rocks of the Western Province.

Cambrian: New Zealand's Oldest Fossils

The oldest known fossils in New Zealand are trilobites from the Cobb Valley in north-west Nelson. They have wonderful names such as *Hypagnostus*, *Kootenia* and *Peronopsis*. They are about 505 million years old, of Middle Cambrian age, and were collected from a formation called the Heath Creek Beds. These may be the oldest rocks in mainland New Zealand.

Cambrian rocks in New Zealand are only known from the Takaka terrane, best thought of as a Cambrian volcanic island arc that was accreted to the margin of Gondwanaland during Early to Late Cambrian time, 530–490 million years ago. The oldest part of New Zealand, the earliest vestige of our modern land mass, was stitched to the continent. Subsequently, younger Ordovician and Silurian sediments were deposited on top.

Cambrian rocks of the Devil River Volcanics are the oldest volcanic rocks known from New Zealand. They were erupted as part of subduction vulcanism at the very margin of Gondwanaland as part of a volcanic arc. They almost certainly formed a chain of volcanoes located in the ocean offshore of Gondwanaland. Interestingly, they are almost identical to rocks of similar age in northern Victoria Land, on the western side of the Ross Sea in Antarctica.

Ordovician Period: New Zealand Moves South

During the Ordovician Period (490–443 million years ago) Gondwanaland swung south, and the New

3. The Cobb River Valley, locality of the breakthrough trilobite fossil find, with the Cobb power station reservoir.

New Zealand's Oldest Macrofossils Discovered by Nelson Schoolboy

In 1948 Malcolm Simpson was a 14-year-old Nelson schoolboy who never imagined he might soon make one of the most important finds in New Zealand geological history – trilobites, the oldest known New Zealand fossils.

Cambrian fossils include trilobites, conodonts, brachiopods, molluscs, sponges and hyolithids. They are known from just a handful of localities in the Cobb Valley, Mt Mytton and Mt Patriarch areas of north-west Nelson. The richest fossil localities are within limestone of the Tasman Formation. This is the oldest limestone known from New Zealand. More than 20 genera of trilobites are present, many of which are yet to be described. No doubt there are many more to be discovered. However, what is known is remarkably similar to fossils of the same age recorded in Australia.

In 1928 Noel Benson of the University of Otago had discovered graptolites (another type of marine animal) in the valley. While holidaying in Nelson in 1948 he decided to visit the Cobb again to see what further discoveries he might make; Simpson, a keen geology student, was invited along for the two-day trip.

Up-valley from the Cobb hydro dam, the party walked until they reached a limestone outcrop, where Simpson hammered off a weathered edge to expose fresh limestone and some indistinct fossils. The cautious Professor Benson, possibly not wanting to come to an on-the-spot conclusion about their significance, described them at the time as 'indistinguishable molluscan remains', but they were later identified and their age confirmed by English geologists.

These fossils – trilobites, now-extinct marine animals related to crabs and woodlice – are New Zealand's oldest macrofossils (fossils that can be seen without a microscope). Vital time indicators, they show that the Cobb Valley limestone is about 505 million years old, making it some of New Zealand's oldest rock.

On the fiftieth anniversary of the find, in 1998, Malcolm was awarded the Wellman Prize by the Geological Society of New Zealand for his contribution to paleontology.

i. As a schoolboy, Malcolm Simpson accompanied the 1948 expedition to the Cobb Valley, inland from Nelson, and discovered fossil trilobites, leading to identification of some of New Zealand's oldest rocks.
ii. One of New Zealand's leading early geologists, Noel Benson first studied graptolites in the 1920s, long before he identified the trilobites in the Cobb Valley.

Zealand component was now at tropical latitudes of 15–20° north.

Ordovician rocks occur in both the Takaka and Buller terranes. They are surprisingly voluminous. Clearly, something big was going on in eastern Gondwanaland to produce this flood of an Ordovician sediments.

Much of the Buller terrane is comprised of greywacke referred to as Greenland Group, and it forms the oldest basement rock of Westland. A single fossil locality is known from the Waitahu River near Reefton, and it is graptolite-bearing. However, there is ample evidence of Ordovician age from other lines of evidence, and in particular the age of detrital minerals: zircons and micas. Greenland Group cannot be older than Ordovician and it cannot be younger.

The Buller terrane lies to the west of the Takaka terrane, and the two are separated by a bounding fault referred to as the Anatoki Thrust. Rocks of the Buller terrane closely resemble those of the Lachlan Fold Belt exposed in Victoria, south-east Australia.

Ordovician fossils are known from rocks in the South Island only, mostly in the north-west Nelson area, but also in southern Fiordland at Chalky Island and Cape Providence. These localities are remote with no road access, but are the only places where Ordovician fossils can be collected on the coast in New Zealand.

Ordovician rocks of the Takaka terrane are limestone, sandstone and shale. The shales (originally mud and silt) in particular are characterised by the widespread occurrence of graptolites.

Other fossils of Ordovician age include corals, sponges, conodonts and trilobites. Fossil sponges are preserved in Takaka Marble that was used as a building stone in some of our finest buildings, including Parliament House in Wellington.

Amazingly, flagstones in the floor of Parliament House have random cross-sections through indistinct and rather poorly preserved fossil sponges. Nevertheless, these must constitute the oldest New Zealand fossils on public display anywhere. How appropriate that they should be in Parliament, one of the oldest institutions in the country.

The age of the black or dark grey variety of Takaka Marble is known: it is of latest Ordovician age, 450–443 million years old (Bolindian).

Conodonts

Conodonts are curious, small, tooth-like structures of primitive fish that resembled hagfish or blind eels. These small fish were extremely abundant and widespread. Long since extinct, they survived for several hundreds of millions of years from Cambrian to Late Triassic time (542–200 million years ago). The actual conodont structures are composed of the mineral apatite and are easily extracted from limestone using acid. A great variety of conodonts are known and they have proven to be immensely useful for determining age.

We know that they belong to these eel-like fish because fossil fish, complete with conodonts in place within their mouths, have been found in Scotland. But until these were discovered and interpreted, the nature of the conodont animal was quite a mystery.

Conodonts in New Zealand rocks (Cambrian to Triassic) were first extracted, prepared and identified by John Simes of GNS Science. He has pioneered a whole field of paleontology in New Zealand, and much of his conodont work has enabled us to establish the age of the older sedimentary rocks of New Zealand.

Graptolites

Graptolites are the fossil remains of strange, delicate, small, stick-like structures only a few centimetres long. Each structure is the skeleton of a small colony of animals, rather like coral. Graptolites were formed by colonial marine organisms that floated in the ocean water mass. Individual animals lived in a small tube or notch set within the bigger structure. Some graptolites look like small hacksaw blades, complete with serrated edges. Others look like leaves, and yet others resemble a bunch of spiders' legs. They come in a bewildering variety of shapes that vary systematically through geological time.

Graptolites are so-named from *graptos*, the Greek word for 'painted or marked with letters', because they form silvery grey or black filmy fossils that resemble pencil sketches or marks on the rock, as if they had been drawn with charcoal on slate. They are very insubstantial, more like impressions, and are best preserved on fine-grained, slaty rocks. Yet they are fantastic for age determination because they evolved fast and spread globally, floating in all oceans. So they occur in marine rocks of Ordovician age almost everywhere.

Although graptolites have long been extinct, there is a rare marine animal called *Rhabdotheca* that exists today, which has a remarkably similar skeleton and life habit. Perhaps this organism is a very distant relative of the graptolites. It has a chitinous or horny skeleton, just like the graptolites would have had.

Graptolites were especially abundant in Ordovician times, but survived until Early Devonian times about 400 million years ago. In all, they existed for about 100 million years, but how they arose and why they disappeared is mysterious.

Graptolite fossils are not especially common in New Zealand, but can be found in rocks exposed in Westhaven Inlet and Cobb Valley in north-west Nelson, and also in the more remote Cape Providence and Preservation Inlet areas of Fiordland.

Graptolite Studies in New Zealand

Graptolites were first recognised in New Zealand by Noel Benson in the 1930s, but were not systematically studied until the 1970s. Then along came Roger Cooper, a New Zealand paleontologist from Wellington. He has described more than 200 species of graptolite from New Zealand Ordovician rocks, all from the South Island. Most of these graptolites are also known from rocks of the same

4. Thanks to the work of GNS Science geologist Roger Cooper we know much more about the complex relationships of New Zealand's oldest rocks. A scientist with many interests, He is the lead author of the 'New Zealand Geological Timescale'. He is best known for his research on graptolites and trilobites.

5. This fossil sponge, seen in cross-section and resembling a large globe artichoke, was originally buried and preserved in limestone on the sea floor and has subsequently been metamorphosed to marble. It is perfectly preserved in this floor tile.

6. The oldest New Zealand fossils on public display in New Zealand are fittingly in Parliament House, Wellington. They can be seen in some of the floor tiles, but only the black ones. They are cut from Takaka Marble (north-west Nelson, South Island) of Ordovician age, between 490 and 443 million years old.

age in south-eastern Australia. In fact, the succession of distinctive graptolite fossils is so well known in New Zealand and Australia that it is now possible to date Ordovician rocks with surprisingly good precision. More than 30 graptolite 'zones' have been recognised.

Silurian and Devonian Periods Leave Few Traces

Not much is known about the history of New Zealand during the Silurian time (443–417 million years ago). It was a period of massive land upheavals and volcanoes. Only two or perhaps three fossil localities are known, and they are all located in a sandstone formation called Hailes Quartzite at Hailes Knob in Takaka Valley, and also the Wangapeka Valley, in north-west Nelson. They are all within Takaka terrane. The fossils are not very diverse but are shelly and include brachiopods, corals and conodonts. They are just distinctive enough to confirm their Silurian age.

During Devonian time, Gondwanaland drifted further into southern latitudes. Although the Devonian Period encompassed a lengthy span (417–359 million years ago), Devonian rocks are almost as rare as Silurian rocks in New Zealand. They are known only from the Reefton area in the Buller terrane, and the Baton River and Skeet River areas in the Takaka terrane of inland northern South Island. The fact that Devonian rocks occur in both terranes is significant and implies that the Buller and Takaka terranes had been accreted and were amalgamated by Devonian times. The Devonian sediments represent a cover that blanketed the new Early Paleozoic margin of Gondwanaland.

The Devonian rocks comprise limestone, sandstone and mudstone. Fossils include brachiopods, bivalves, gastropods, corals, sponges, bryozoans and conodonts. Devonian limestone near Reefton is famous for its brachiopod shellbeds with *Acrospirifer*, described by Robin Allan of the University of Canterbury. Yet many brachiopods remain to be described.

A rich and diverse bivalve assemblage has been described by Margaret Bradshaw, also at the University of Canterbury, and it includes 27 genera and 46 species. This is a very significant contribution to our knowledge of ancient New Zealand. It is pretty much the oldest molluscan fauna known from New Zealand, the oldest shellbeds. Very sparse molluscan fossils are known from our Cambrian and Ordovician rocks, but nothing so wonderful as the Devonian.

Also of great interest is the occurrence of the oldest bryozoans known from New Zealand. Bear in mind that by far the great bulk of our young Cenozoic limestones (less than 65 million years old) is comprised of bryozoans. They are serious rock formers!

Ancient Fish Appear

Late Devonian time saw the first appearance of the ancestors of 'modern' fish: the sharks, rays and bony fishes such as salmon, snapper and eel. A number of fossil fish have been recovered from the Waitahu Valley near Reefton. They include species of at least two forms: placoderms ('plate-skinned') and acanthodians ('spine-finned'). Conodonts aside, these are the oldest fish fossils known from New Zealand.

The armour-plated *Actinolepis* (placoderm) may have grown up to 2 metres long, whereas *Nostolepis* and *Gomphoncus* (acanthodian) species were small, about 20 centimetres in length.

7. Resembling hieroglyphs, graptolites were marine organisms common in the ocean to about 400 million years ago. And like hieroglyphs, they have enabled scientists who have learned to 'read' them to unlock secrets of the past. At least three different forms of graptolite are preserved on this rock surface, including *Isograptus*, *Tetragraptus* and *Pseudisograptus*.

8. Until about 275 million years ago, brachiopods, or lampshells, used to be the world's dominant marine shellfish. They can still be found in the waters around New Zealand today. Packed with fossils of the brachiopod *Acrospirifer coxi*, this shellbed was discovered in limestone near Reefton and is of Devonian age (417–359 million years old).

7

8

07/ Hot Rocks: Median Batholith

Median Batholith

Between the basement rocks of the Western Province and Eastern Province there is an extraordinary belt of primarily igneous rocks, characterised by plutonic rocks that range in age from Late Devonian to Early Cretaceous (385–99 million years ago). They represent a 260-million-year-old locus of magmatism and would have been the main source of volcanism in the New Zealand sector of eastern Gondwanaland.

The Median Batholith is well exposed in the western South Island, especially north-west Nelson, West Coast, Fiordland and Stewart Island, and must extend northwards beneath younger rocks of western North Island. From all available geological observations, it must have developed *in situ* adjacent to the Western Province.

It entirely predates the separation of Zealandia from Gondwanaland and yet was the locus of Early Cretaceous granite emplacement associated with the rifting of Zealandia. It includes many plutonic and metamorphic rocks, the ages of which have all been determined by radiometric dating over the past 40 years. There are almost no sedimentary rocks and no known fossils associated with the Median Batholith. This is hardly surprising given that it is

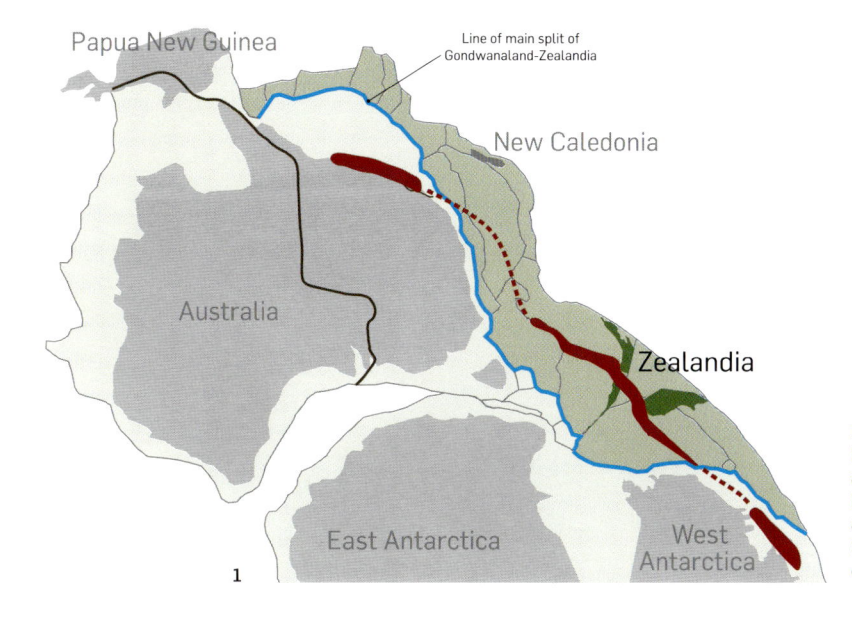

1. On this reconstruction, the Median Batholith is shown as an elongate belt dominated by granites of Cretaceous age, and stretching for 3000 km along the margin of eastern Gondwanaland.

overwhelmingly crystalline, in other words dominated by igneous and metamorphic rocks that have cooled from temperatures well in excess of 250°C. The granites, of course, have all cooled from magma at temperatures of about 850°C. Any sediments or sedimentary rocks that may have had fossils have been well and truly cooked so that any organic material has been totally destroyed.

The concept of a long-lived batholith is a relatively new development in our understanding of New Zealand's geology and is largely the brainchild of Nick Mortimer and Andy Tulloch, both at GNS Science based in Dunedin. They have recognised the full extent of the Mediam Batholith, stretching more than 3000 kilometres along the margin of what was once eastern Gondwanaland.

Over the past decade, there has been an intensive mapping campaign in both Stewart Island and Fiordland, and this work has generated a vast amount of new information about these remote areas of New Zealand. All of it conforms to this concept of a Median Batholith.

Hot Tourism

The Median Batholith would make an ideal tourist destination. But if so, the North Island would miss out! It all occurs within some of the South Island and much of it lies within New Zealand's finest recreational and wilderness areas, such as the Abel Tasman, Kahurangi and Fiordland national parks, and Stewart Island. There are some

2. Nick Mortimer (GNS Science) is a petrologist and one of New Zealand's foremost authorities on the nature, origins and relationships of the basement rocks of New Zealand.

3. Andy Tulloch (GNS Science) is also a petrologist and one of New Zealand's authorities on granite.

4. Granite plutons and batholiths are largely confined to the South Island and Stewart Island. They form a distinctive belt of igneous rocks that relate to the ancient site of rifting within eastern Gondwanaland that gave rise to Zealandia. They represent the rupture of continental crust and the rise of new continental crust (fresh cream). The bulk of these granites are between 125 and 105 million years old.

5. Much of Stewart Island is mapped as part of the Median Batholith. It is famous for its granite, and none more so than here at fragile Gog Magog at the southern end of the island. Sculpted by the ravages of freeze and thaw, the granite here is so weathered that it can be shovelled. Yet it is of Cretaceous age, between 125 and 105 million years old, representative of the rifting of continental crust within eastern Gondwanaland immediately prior to the formation of the Tasman Sea floor.

4

well-known tourist spots, such as the Heaphy Track, Karamea, Kaiteriteri and the coast between Motueka and Takaka, the Victoria and Brunner ranges near Reefton, not to mention the Homer Tunnel, Milford Sound and almost all of the rest of Fiordland.

These rocks all formed at depth within the crust but are now revealed and exposed at the surface because of tectonic uplift and subsequent erosion.

Stewart Island

Detailed geological mapping of Stewart Island has revealed a large number of individual plutons, each derived from a single batch of magma or several closely related magmas. Needless to say they have all been named and many of them have been dated; in other words, their age of crystallisation has been determined. Some of the 38 formally named plutons are listed here, and all of them relate to topographic features or areas on Stewart Island: Ruggedy, Neck, Knob, Freds Camp, Big Glory, Forked, Euchre, Codfish, Deceit, Rollers, Tarpaulin, Smoky, Easy, Tikotitahi, Doughboy, Blaikies, Gog, Kanihinihini and Lords.

Each pluton varies in visible dimension and volume but is generally homogeneous, more or less the same in consistency (composition, colour, crystal size, texture and structure) throughout the pluton body. They can be thought of as originating as large bodies of hot liquid rock extruded from within the mantle or lower crust and cooled at some higher level within the crust, like great globs of glue. They probably behave like ascending bodies of hot wax in a lava lamp, with larger distorted globular to mushroom shapes.

Easy to Date

In general, the rocks of the Median Batholith are easy to date. The rocks are loaded with suitable quantities of beautifully crystalline uranium- and potassium-bearing minerals. The granite plutons on Stewart Island are mainly of Carboniferous, Jurassic or Cretaceous age, based on dating of zircons and micas.

Granites and Gabbros

Plutonic rocks vary considerably in appearance, and especially in crystal size (or grain-size) and colour. It all depends on their silicon composition. If it is similar to basalt (less than 52 percent), it will be dark, and will be referred to as gabbro if coarse-grained, or dolerite if medium-grained. In fact, gabbro may be thought of as a plutonic form of basalt.

Bluff Hill near Bluff is famous for its handsome gabbro and dolerite, of Triassic age. Most people are unfamiliar with this rock term, and yet it is widely used in the building industry and as a monumental stone, especially for headstones on graves.

5

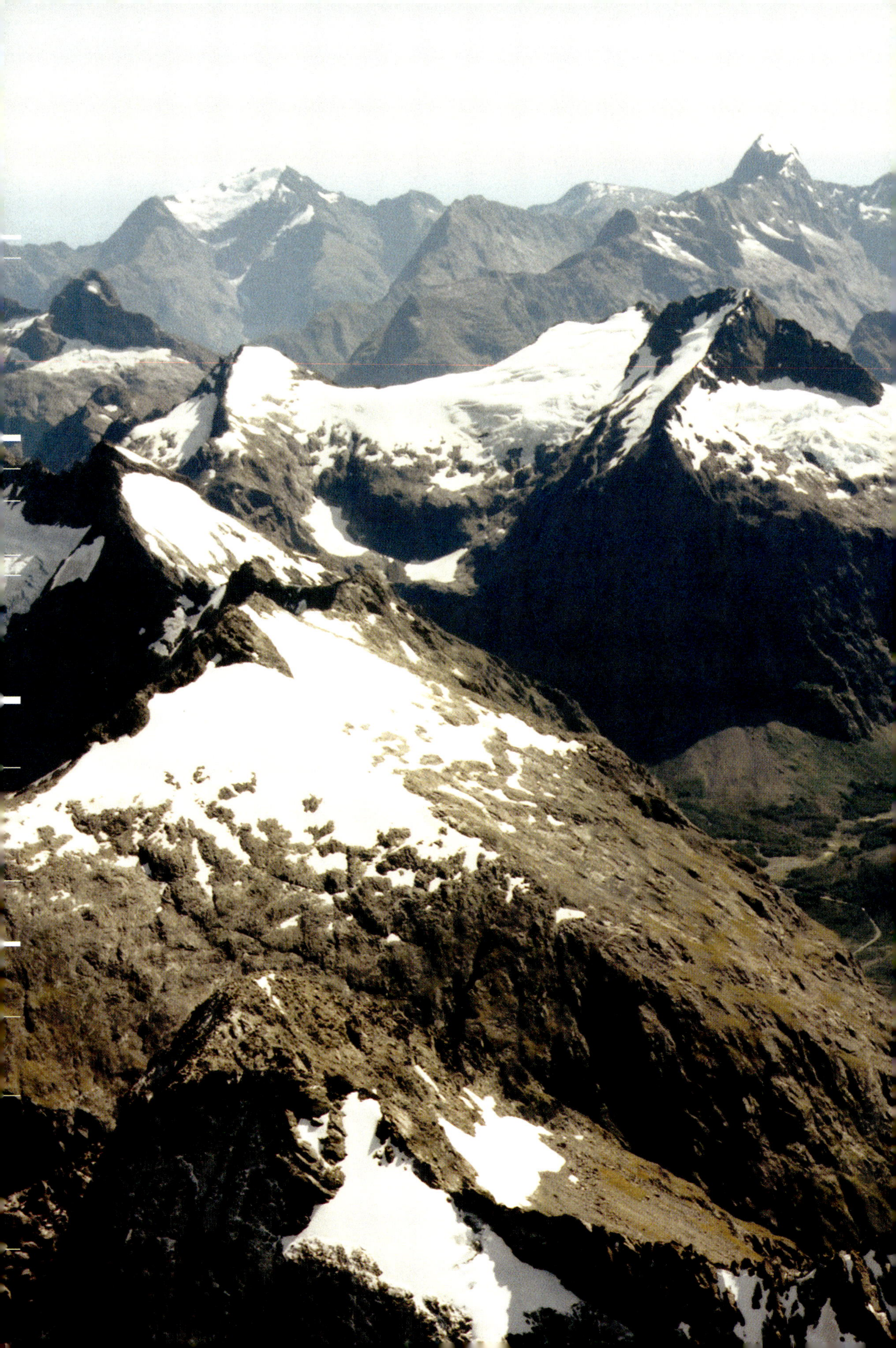

6. Much of Fiordland is within the Median Batholith. Some of New Zealand's most magnificent mountain scenery owes its character to the strength of these amazingly beautiful crystalline igneous rock formations such as the Western Fiordland Orthogneiss and the Darran Diorite. Some of these rocks are derived from very considerable depths within the Earth's crust, but the rapid uplift of Fiordland has brought them to the surface. This view is looking north up the Hollyford Valley from near the Homer Tunnel on the road to Milford Sound. Scalloped by repeated glaciations during the past 2 million years, these tough rocks hold their shape well, preserving their ice-cut heritage. The highest point in Fiordland, Mt Tutoko, looms above all else.

Gneiss . . . As In Nice

Granites and gabbros are igneous, derived from molten rock or magma, but they are not the only crystalline rocks within the Median Batholith. Some are metamorphic. As stated earlier, any pre-existing rock can be metamorphosed or recrystallised. Coarsely crystalline or coarse-grained metamorphic rocks are referred to as gneiss, yet another rock term of Scandinavian and/or German origin.

Furthermore, it is possible to distinguish between gneiss that is metamorphosed granite (or plutonic rock), and a metamorphosed sedimentary or metamorphic rock (such as schist). The former is referred to as orthogneiss and the latter as paragneiss.

And there is no better example of an orthogneiss as the Western Fiordland Orthogneiss. This is the rock that hosts the Homer Tunnel and from which some of Fiordland's most spectacular mountains are sculpted, including Tutoko, its highest peak, and Mitre Peak, its most picturesque peak.

A Mystery Solved?

One of the great mysteries in New Zealand geology is the whereabouts of the volcanoes that produced all the sediment within the Murihiku terrane (Eastern Province). These rocks form an elongate belt that can be traced the length of New Zealand, but is most voluminous in Southland where it forms a great tract of country, everything south of a line drawn between Te Anau and Balclutha, and east of a line drawn between Te Anau and Riverton. These rocks are almost exclusively made up of sediments that are derived from subduction-related volcanoes, and yet the actual volcanoes are no longer evident. All that remains is a great stack of sedimentary rocks that appears to be greater than 12 kilometres thick! These are the rocks that make up the so-called Southland Syncline, a great folded pile of sediments.

The most plausible location of the volcanoes is the Median Batholith, and indeed suitable plutonic igneous rocks that may have fuelled the magma chambers that erupted as volcanoes during Late Permian through to Late Jurassic time have now been found. They are not exactly volcanoes any longer, but this is hardly surprising. Erosion has set in big-time, and all that can now be seen are the deeper remnants of the magmatic system.

Karamea: Perhaps the Oldest Pluton in the Median Batholith

Outside Te Papa there are three large rocks symbolising the land of New Zealand, the original people who settled here (Maori), and all those people who have arrived since, subject to the Treaty of Waitangi. This latter rock therefore relates to *tangata tiriti*. Whereas the first two rocks are conceptually indistinguishable one from the other (*papatuanuku* and *tangata whenua*) and are therefore of the same rock type (andesite from Mt Taranaki), the third rock is very different. It is a piece of Karamea Granite and is of Devonian age, 375 million years old.

Karamea Granite is a spectacular pink, white and black granite with large crystals of feldspar, quartz and biotite mica, respectively. Ironically it looks angular, fresh and youthful compared to the older-looking, smooth, rounded but drab, grey, fine-grained andesite boulders. No doubt this is deliberate symbolism, carefully chosen. The granite is from a quarry within Karamea Granite in the Oparara River on the West Coast.

Orbicular Granite

One of the more bizarre granites ever found in New Zealand is the famous orbicular granite boulder found near Karamea early in the twentieth century. So unusual by virtue of its curious nested coloured orb or ring structure, it was laboriously retrieved and cut into a series of parallel-sided slabs.

Patrick Marshall was responsible for this. The slabs were then distributed to a number of scientific institutions, including some overseas, such as Cambridge University in England, as curios. Te Papa has one on display. GNS Science has another, as do the universities of Otago and Auckland. The strange thing is that the actual source of this distinctive rock type has not yet been established. It is presumed to be from somewhere within the Karamea Granite, but exactly where remains a mystery to this day.

Beyond the South Island
The Median Batholith is recognised beyond the New Zealand land mass from geophysical surveying, and especially seismic interpretation. Furthermore, granites do occur on Auckland Island and the Bounty Islands, far-flung emergent outposts of the Zealandian continent. They provide valuable insight into the largely unknown Campbell Plateau and confirmation of the extent of continental crust of Zealandia. They have been dated and are entirely compatible with the Median Batholith.

The Ultimate Mystery: Source of the Greywacke?
The Median Batholith informs us of the fiery origins of a fair swathe of New Zealand's crust. Critical to this understanding is our ability to date the rocks. Their age relationships are all important, as is their distribution based on mapping. At present, the Fiordland area is subject to detailed mapping and sampling as part of the GNS Science QMap project. Not for the faint-hearted! Fiordland is exceptionally difficult terrain to map. Yet the result will be similar to what has been accomplished within the past decade in Stewart Island: a much greater knowledge of the geology of the contents of the Median Batholith and its history.

Curiously, the Median Batholith is not the source of the sediments that make up the great bulk of the Eastern Province, everything that lies to the east of the Median Batholith. This is the greywacke and schist rock that forms such a large part of the New Zealand land mass today. For all that they are indeed derived from granitic rocks, it can be said categorically and without hesitation that the granites of the Median Batholith are not the source. There are strong reasons for this. For one, much of the granites that make up the Median Batholith are of Cretaceous age (145–65 million years old) and are therefore much younger than the greywacke. Read on! The next chapter deals with the Eastern Province.

Is It Dead Yet?
One might ask this question. If the Median Batholith was the locus of magmatic activity and volcanism for 260 million years, could it still be live? The youngest rocks are about 100 million years old. Say no more! It seems that it is indeed dead.

08/ The Younger Old Rocks: Eastern Province

Eastern Province: Rocky Latecomers

Between 310 and 125 million years ago, a span of 183 million years, the basement rocks of Eastern Province were laid down. Virtually all sedimentary, these Eastern Province rocks are comprised of seven terranes and lie entirely east of the Median Batholith. From west to east they are named as follows: Brook Street, Murihiku, Dun Mountain-Maitai, Caples, Waipapa, Rakaia and Pahau. These terranes were successively added to the margin of Gondwanaland and were all docked, so to speak, in the Early Cretaceous, about 125 million years ago.

This assembly of terranes was then buried by a blanket of younger sedimentary rocks. Initially they were marine sediments – that is, they accumulated in the sea along the eastern margin of Gondwanaland. Then they were pushed up to become land and the subsequent cover sediments became terrestrial (non-marine), just like the sediments associated with modern-day land surfaces, such as sand dunes, soils, swamps, lakes and river deposits.

Let us first think of the history of the Eastern Province through geological time, starting with the Carboniferous.

1. Conodonts: tiny phosphatic, cone-shaped, teeth-like fossils that preserve well in marine sedimentary rocks. They are 'jaw apparatus' from primitive fish that no longer exist but which may have been similar to the agnathids (hagfish). They are known from rocks of Cambrian to Triassic age. No Jurassic conodonts have ever been found. Because they are phosphatic, they are easily recovered from rocks that can be dissolved in acid such as limestone.

Carboniferous: Where Have All the Rocks and Fossils Gone?

During the Carboniferous Period (359–299 million years ago) plant life blossomed and, in decaying, these plants became peat, and ultimately coal. Hence the term 'Carboniferous'. There are extensive coals of this age in Europe and the USA. No New Zealand coals were created at this time, however. That would happen later, 100–37 million years ago.

Even more rare than Silurian and Devonian rocks and fossils, Carboniferous rocks and fossils are almost unknown in New Zealand. This is rather mysterious and has New Zealand geologists puzzled. Where have all the Carboniferous rocks gone? Perhaps there are significant areas of Carboniferous rocks yet to be discovered, buried out of sight on the Campbell Plateau within submerged Zealandia. They must be somewhere! We can be certain that

New Zealand existed during this major period of the Earth's history and therefore there must be a record. We have just not found it yet.

There are indeed some Carboniferous rocks in New Zealand, but not many. There are three fossil localities known, all from isolated limestones in Rakaia terrane rocks of north Otago and south Canterbury. The fossils are conodonts of variable preservation. Because Carboniferous fossils and rocks are almost non-existent in New Zealand, these localities have assumed an inordinate importance. The most famous of them is at Kakahu, near Geraldine. Two paleontologists, the brothers Hugh and Graeme Jenkins, discovered conodonts and fossil fish scales here in the late 1960s.

A possible fourth locality in Waipapa terrane rocks at Purerua Peninsula in the Bay of Islands may contain foraminifera (tiny fossils of marine animals) of Carboniferous age. If this were verified, then this occurrence would represent the oldest fossils and rocks known from the North Island.

Permian Period: New Zealand Moves Deep South

For much of the Permian Period (299–251 million years ago), the New Zealand part of Gondwanaland was underwater and situated closer to the South Pole, at 60°–70° latitude. Volcanic islands popped their peaks above the sea and undersea New Zealand became a dumping ground for material sediment derived from Antarctica and Australia.

Permian shelly fossils are known from at least three localities in Waipapa terrane rocks of Northland (Moturoa Island, Bay of Islands; Arrow Rocks and Marble Bay, Whangaroa Bay) in the North Island, and many localities in the South Island: the Nelson area, north-west Nelson (Parapara Peak), Southland, west Otago and south Canterbury. Permian fossils occur in the Brook Street, Murihiku, Dun Mountain-Maitai, Caples and Rakaia terranes. So there is a widespread Permian record in the Eastern Province, but by far the best fossil record is in the Brook Street terrane (curiously named after a street in Nelson).

Permian fossil localities are nevertheless not especially common and are also hard to access. Most of them are on private land and, apart from a few places on D'Urville Island, there is no coastal exposure of Permian fossil-bearing rocks. In the Nelson area, the Maitai and Wairoa Rivers offer the best chances of fossil collecting; likewise, in Southland the Wairaki River that drains the Takitimu Mountains and Wairaki Hills. The richest Permian fossil locality is in Productus Creek, a small tributary of the Wairaki River.

Fossils include brachiopods in particular, but also molluscs (bivalves, gastropods, rostroconchs, scaphopods, cephalopods), corals, bryozoans, echinoderms, sponges, trilobites, conulariids and fish scales. Tiny fossils that are only visible with the aid of a microscope have also been recorded, such as foraminifera, conodonts and radiolarians.

The cephalopods include the nautiloids and ammonoids, which are great fossils to find because they generally preserve well and are fascinating objects to look at. However, as yet almost no Permian cephalopods are known from New Zealand. Very small ammonoids have been collected from limestone in Rakaia terrane rocks in hills above Lake Aviemore, and only two straight orthocones have been found in Brook Street terrane rocks in the Wairaki Hills. The absence of ammonoids from our Permian shellbeds is a little puzzling and also very disappointing, because they are so useful for determining age.

A few trilobites of Late Permian age have been collected from two localities in Southland. These represent some of the youngest trilobites known globally. They finally became extinct at the end of the Permian.

Not a single Permian bone fossil is known from New Zealand. However, some fossil fish scales have been discovered and even identified as possibly belonging to ancestral gar fish.

Conulariids

Conulariids are curious, elongate, tapering cones no longer than 10 centimetres

2. The Takitimu Mountains in Southland are an isolated, canoe-shaped range, named after one of the original Maori canoes. They are the remnants of a great pile of lava flows and volcanic sediments. These rocks accumulated in the sea adjacent to active volcanoes that must have existed as part of a subduction-related island arc off the coast of Gondwanaland. Fossils tell us that this chain of island volcanoes (rather like Vanuatu today) erupted during much of Permian time. This view is looking north from Letham Ridge in the Wairaki Hills, towards Mt Hamilton (extreme right distance). The Takitimus are ringed by extensive native beech forest with clearly defined upper and lower tree lines. The valley is within the catchment of the Wairaki River.

3. *Paraconularia*, a strange, tapering, conical fossil of as yet unclear biological affinity. Nothing quite like this exists today. Yet the conulariids were around for more than 300 million years. This handsome specimen is of latest Triassic age and must represent one of the youngest conulariids ever found. It was actually found in New Caledonia, but identical fossils are known from Southland. The 'shell' is phosphatic and it is thought that the animal was a coelenterate, a distant relative of the sponges and corals.

4. Bruce Waterhouse (Oamaru; formerly Dorothy Hill Professor of Paleontology, University of Queensland), an authority on Permian fossils and rocks of the world, has described most of the Permian shelly fossils (brachiopods and molluscs) of New Zealand.

long, with box-like cross-section. They have a long age range, becoming extinct at the end of the Triassic, but what kind of animal they are remains a mystery. They are thought to be some sort of strange coelenterate or sponge-like creature. New Zealand has some of the youngest conulariids known globally, from rocks about 205 million years old. Conulariids of the same age are also found in New Caledonia where rocks of Late Triassic age are almost identical to those of New Zealand.

Permian Fossils Confirm Gondwanaland Origins

Permian fossils from New Zealand have been studied almost exclusively by Bruce Waterhouse since the late 1950s. He has systematically described almost all known fossil groups. What he has established is that they are most like those of eastern Australia. This is a very satisfactory result and fits with our story of the systematic growth of Gondwanaland.

Brachiopods were especially abundant and diverse in the Permian, but the molluscs were on the rise. Of particular note is the widespread occurrence of a distinctive bivalve group (named the 'atomodesmatinids') with a curious shell structure. The outer layer is comprised of tiny, elongate prisms of calcite, just like modern-day horse mussels (*Pinna*). This type of shell is quite unusual, yet it is characteristic of moderate to large clams, referred to collectively as 'inoceramids', in the Permian, Jurassic and Cretaceous. For some reason they were not conspicuous in the Triassic. They form shellbeds and limestones.

Rostroconchs are primitive molluscs with two valves rather like a bivalve, but the valves are not attached. Unlike bivalves, they do not seem to have been connected by a ligament along their hinge. Rostroconchs range in age from Cambrian to Late Permian, so they lasted several hundreds of millions of years (542–251 million years ago).

The Amazing Atomodesmatinid Clams

Most notably in New Zealand, there are limestones that appear to be entirely comprised of calcite prisms derived from these particular bivalves (atomodesmatinids). The Wooded Peak Limestone of the Nelson area is hundreds of metres thick, becoming up to 1000 metres thick in places. Amazingly, this formation can be traced through the South Island for almost 200 kilometres. It is extremely hard to imagine the kind of marine environmental conditions that enabled these particular shells to grow in such abundance, die and then disintegrate, only to accumulate in such a manner as to produce such a vast volume of almost pure shell sand.

New Zealand's Oldest Plants: *Glossopteris*

The oldest plant fossils known from New Zealand are of Late Permian age and they include leaf fossils of *Glossopteris*. However, only about 10 specimens have ever been collected. *Glossopteris*, sporting thick and fleshy, tongue-like leaves, belongs to an immensely important group of extinct plants that dominated the forests of Gondwanaland. The presence of *Glossopteris* is regarded as a defining characteristic that is synonymous with Gondwanaland. When the first *Glossopteris* leaf from New Zealand was described and identified by Dallas Mildenhall (GNS Science) in 1970, it was a red-letter day! New Zealand could at last claim to have been part of Gondwanaland. Without the proof of this one fossil, there was no certainty that it had. As Swiss geology Professor Rudolf Trümpy once said: 'A bad fossil is more valuable than a working hypothesis.'

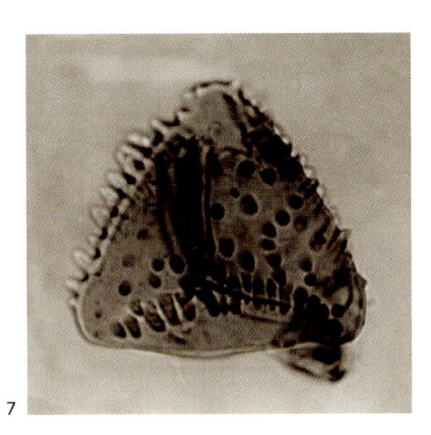

5. A distinctive fossil leaf of the tree family that dominated Gondwanaland and has come to symbolise it: *Glossopteris*. First recognised in New Zealand in 1970 by Dallas Mildenhall, at least 10 fragmental leaf fossils are now known (not many, but enough!) and all are from rock formations exposed in Productus Creek.

6. Dallas Mildenhall (GNS Science), one of New Zealand's few pollen experts (a palynologist) with special interests in the history and origins of the New Zealand flora, floral change during the ice ages (Pleistocene; the past 1.8 million years), and a leading forensic scientist.

7. One of the oldest known fossil spores from New Zealand. Using powerful acids, this has been recovered from rocks of Late Permian age exposed in Productus Creek in the Wairaki Hills on the eastern side of the Takitimu Mountains. Fossil pollen and spores are more resilient than rock! This photograph was taken down a microscope under strong magnification.

Some fossil wood has been found, but remains as yet unstudied. All these occurrences are known from Southland and in particular the Wairaki Hills just north of Ohai. These rocks are all within the Brook Street terrane.

The oldest palynomorphs are also of Late Permian age. These are fossil spores and pollen. Oddly, the best-preserved Permian palynomorphs from New Zealand have been recovered from much younger Cretaceous sediments in the Chatham Islands. This highlights one of the problems of paleontology: the reworking of fossils, especially microfossils. As a rule, it is easy to identify and distinguish reworked fossils, but, if the preservation of the recycled material is much better than the non-recycled fossils, wrong interpretations can be made.

Triassic: Enter the Dinosaurs and Mammals

The Triassic age (251–200 million years ago) was an important time in the Earth's history, not least because the first dinosaurs and mammals appeared, about 225 million years ago. However, there are as yet no Triassic fossils known from New Zealand that can be attributed to either dinosaurs or mammals.

Triassic fossils were first recognised from New Zealand by Ferdinand von Hochstetter while visiting the Nelson area in 1859. An outstanding geologist, he came here as a natural scientist with the *Novara* on a grand voyage of discovery, sent out by the Austro-Hungarian Empire. While here, Hochstetter was commissioned by the Nelson Provincial Government to make observations on the geology of Nelson. He did a fine job and produced the first geological map of the Nelson region, an extraordinary achievement all done within a few months.

Hochstetter must have been amazed to find fossils which he was familiar with from the Austrian Alps, and in particular the bivalves *Monotis* and *Halobia*. He found them as he traversed a relatively steep slope on the Barnicoat Range immediately behind Richmond. He was able to identify them correctly and determine their age as Late Triassic. But these were just two species of many that he collected.

On his return to Austria, he so impressed the Emperor Franz Josef that he was able to house his New Zealand collections in a palace, the Natural History Museum in Vienna. To this day, these precious first collections from Nelson are housed in unparalleled splendour. Furthermore, there are magnificent scenes of New Zealand painted high up the walls of at least one room in the museum. Every New Zealander interested in natural history should pay a visit.

Triassic (in fact Late Triassic) sedimentary rocks form the greater part of the basement rocks of New Zealand. They vie with the Ordovician for top spot in the rock dominance stakes. It could be said that the New Zealand land mass is substantially comprised of Ordovician and Triassic rocks. It is thought that much of submerged Zealandia of the Campbell Plateau is probably of Ordovician age.

Triassic rocks occur in five of the seven Eastern Province terranes and are especially voluminous in the Murihiku, Dun Mountain-Maitai, Caples and Rakaia terranes.

Just as in the Ordovician Period, something significant must have happened in Gondwanaland to produce such a vast flood of sediment of Triassic age. In order to understand the process, scientists have looked at the large fans of sandstone and mudstone on the sea floor below the deltas of large rivers such as the Indus, Ganges and Amazon. Over time, sand from rivers builds up on the continental shelf, then pours down slopes into ever-deeper water, transported in submarine slides. Meanwhile, on land, an uplift of mountainous terrane is required with high rainfall to generate the flood of sand that is represented by the Triassic greywacke rocks of New Zealand.

Triassic Starts With a Bang

The Triassic Period commenced 251 million years ago, with a massive extinction. The end of the Permian or Permian–Triassic boundary event had a bigger impact on life than any other in the Earth's history. It is estimated that perhaps as much as 90 percent of all life was eliminated.

The cause of this event remains unknown and continues to be debated. There is a suggestion that the cause was the largest volcanic eruption in Earth's history – the Siberian Traps in Russia. The volume of lava is estimated as 1–4 million cubic kilometres, much greater than the largest twentieth century eruption, that of Mt Pinatubo in 1991–2, which caused a 0.5°C drop in global temperatures. Certainly the Siberian Traps are extremely voluminous basalts, the ultimate LIP, and they are the right age to support this theory. To put this in context, all of Auckland's volcanoes amount to less than 3.5 cubic kilometres!

Or Was it a Comet?

There is evidence of an extraterrestrial source such as a comet impact. Most compelling is the tell-tale presence of strange crystalline carbon molecules called 'buckminsterfullerines'. These crystals have been reported from a distinctive 'boundary clay' in China and are considered to be alien to any known terrestrial process. The only way they could form is from an extreme, high-pressure impact, such as might be generated by a comet or a meteorite hit.

A comet impact is the preferred theory, because comets travel much faster than meteorites, are mainly comprised of water and carbon dioxide, make an incredibly devastating mess on impact, and yet leave almost no calling card. Comet impacts are, therefore, very different from meteorite impacts, such as that which formed the Chicxulub Crater responsible for the less damaging but more famous Cretaceous–Paleogene boundary extinction event (65 million years ago). Tell-tale chemical signatures loaded with unusual concentrations of rare metals such as iridium have been linked with the Chicxulub meteorite impact, but there is nothing like this associated with the Permian–Triassic boundary. A dirty snowball comet would leave no metallic signature: just water and carbon dioxide.

In New Zealand, we know of at least three places where the Permian–Triassic boundary may be found. Obviously, they all have rocks of Late Permian and Early Triassic age, but pinpointing the precise boundary is not easy. There are no natural signposts saying: 'This is it: the P–T boundary.'

Parapara Peak

One of these localities is in the Western Province at Parapara Peak in north-west Nelson. This is the least hopeful of the three, because the rocks have been significantly metamorphosed and altered and the prospect of finding fossils or traces of any organic matter that might preserve a record of the chemistry of the sea that prevailed at the time the sediments were deposited on the sea floor, are negligible. The so-called 'stable isotopes' of carbon and oxygen are routinely analysed in fossil organic matter and provide useful information on physical aspects of the ocean, such as water temperature.

Arrow Rocks

An excellent Late Permian–Early Triassic rock sequence is exposed on Arrow Rocks, a tiny islet less than a kilometre offshore of Marble Bay in Whangaroa Bay on the eastern Pacific Ocean side of Northland. The rocks here are part of the Waipapa terrane and are sedimentary rocks that accumulated in a deep marine setting, in at least 2 kilometres of water depth. Geologists are still working on establishing the precise location of the Permian–Triassic boundary in these rocks.

The rocks contain abundant microfossils: radiolarians and conodonts. Geologists exclusively study the fossils of the radiolarians here, which are tiny single-celled animals that live in seawater. They produce exquisitely ornate basket-like skeletons from silica (silicon dioxide) and are between 30 and 100 microns in size (1 micron is 1000th of a millimetre or one millionth of a metre)! The name 'radiolarian' relates to the structure of their skeletons in that they resemble radio (or television) transmission towers. These tiny animals are abundant in the ocean today and have been since Cambrian time. They have a very long history, more than 500 million years, and form a significant component of plankton at the very base of the food chain. The skeletons of these tiny animals form a rain of biogenic sediment in parts of the open ocean floor, and these sediments, known as radiolarian oozes, form distinctive rocks known as chert.

A long-term, collaborative research project on Arrow Rocks has been undertaken involving eight Japanese universities, the University of Auckland and GNS Science. The Japanese leaders are Yoshiaki Aita, Atsushi Takemura and Rie Hori, and they are all radiolarian paleontologists. Their work is painstaking and laborious and the strike rate is low: for every 100 samples only two will be productive. On Arrow Rocks there are about 150 metres of stratified rock, and much of it is in ultra-thin layers. The preservation of microfossils is highly unpredictable and not apparent until the sample is processed in the laboratory.

The actual fossils are tiny, measurable in microns.

The impact of radiolarian paleontology on our understanding of New Zealand basement geology cannot be overstated. Huge advances in our knowledge have been made since the mid-1980s, with the arrival of Yoshiaki Aita as a post-doctoral fellow at the University of Auckland. He was attracted to New Zealand by Bernhard Spörli, one of our foremost structural geologists. Prior to this breakthrough in approach, the old sedimentary basement rocks of New Zealand seemed intractable and uninviting. It was all too hard and few geologists could muster any interest in these rocks, but this has all changed.

Maitai Group

The third Permian–Triassic boundary sequence is within Maitai Group rocks of the Dun Mountain-Maitai terrane and is different again. It can be seen in many places in both Nelson and Southland and is always associated with the same two formations: Permian Tramway Formation and the overlying Little Ben Sandstone (the names of these formations relate to geographic features in the Nelson area). The precise location of the boundary remains obscure but may yet be discovered. The original sediments are thought to have accumulated in water depths of less than 200 metres, but even this is uncertain. Carbon isotopic studies suggest that the Permian–Triassic boundary is located within the basal part of the Little Ben Sandstone.

Early Triassic

There are not many fossils of Early Triassic age (251–245 million years ago) known from New Zealand, but there is a lot of rock that must be of this age, especially in the Caples and Dun Mountain–Maitai terranes. The best-known Early Triassic fossils include conodonts and radiolarians from Arrow Rocks (Waipapa terrane), and ammonoids from the Nelson area (Dun Mountain–Maitai terrane) in the Wairoa River and on D'Urville Island. Ammonoid fossils have been recorded or described by Bruce Waterhouse and Stuart Owen (formerly of Otago University).

The most accessible locality in New Zealand with fossil-bearing rocks of Early Triassic age is at Kaka Point, south of Balclutha on the south Otago coast. Public access is easy – the coast from Kaka Point to Nugget Point affords access to the oldest fossils that are easily reached in New Zealand. However, the fossils are not

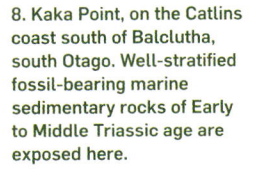

8. Kaka Point, on the Catlins coast south of Balclutha, south Otago. Well-stratified fossil-bearing marine sedimentary rocks of Early to Middle Triassic age are exposed here.

9. A superb specimen of the ammonoid fossil *Simplicites*, yet typical of ammonoids of Early to Middle Triassic age collected from rocks exposed on the shore platform at Kaka Point. You can just make out fragments of the ceratitic suture pattern preserved on the outer whorl.

10. Arrow Rocks, Whangaroa Bay, Northland. This small islet has been subject to intense research by a dedicated team of Japanese and New Zealand paleontologists interested in the history of the Panthalassa Ocean during Late Permian to Late Triassic time, and in particular the history of radiolarian plankton across the Permian–Triassic boundary. On location at the southern tip of the islet, from left to right: Satoshi Yamakita, Nori Suzuki, a student, Yoshiaki Aita and Atsushi Takemura.

11. Doug Campbell (University of Otago), with specialist interests in the greywacke rocks of New Zealand, Triassic paleontology and stratigraphy, fossil brachiopods and paleobotany.

12. New Zealand's oldest known terrestrial animal fossil: a single bone from the shoulder apparatus of a distant cousin of the frogs, a Middle Triassic stereospondyl labyrinthodont amphibian. A: The actual fossil, collected from a small quarry on a farm due east of Invercargill in a tributary of the Mataura River.

B: A sketch showing the location of the bone within a 2–3 m salamander-like amphibian. These animals were abundant globally in Carboniferous to Triassic time prior to the rise of the dinosaurs and mammals in Late Triassic time.

12A 12B

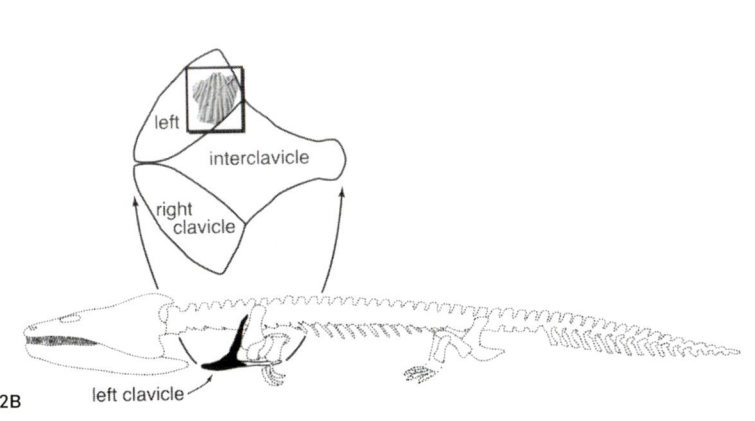

exactly conspicuous; they are microfossils of plankton. Radiolarians have been recovered from phosphatic nodules collected from siltstones by Rie Hori.

Oldest Marine Reptile
The oldest marine reptile fossil from New Zealand, a nothosaur, is from Early Triassic rocks at Rocky Point near Mossburn. This fossil was excavated in the early 1990s by Ewan Fordyce (University of Otago), and required temporary closure of the main road to Te Anau. A remarkably complete skeleton was recovered, but unfortunately the skull was missing. This find is about 248 million years old, based on associated radiolarian fossils identified by Yoshiaki Aita.

Middle Triassic
Rocks and fossils of Middle Triassic age (245–237 million years ago) are much richer and much more diverse, and include brachiopod shellbeds. They are especially widespread in Southland and Nelson (Murihiku terrane) and can be found to a lesser extent in Otago and Canterbury (Rakaia terrane). Famous localities of Middle Triassic age in Canterbury are known from the Mt Potts, Otematata and Harper Range areas. Curiously, the only fossils of Middle Triassic age known from the North Island are conodonts and radiolarians in Northland (Waipapa terrane), at Arrow Rocks and Mahinepua Peninsula, Whangaroa Bay.

Oldest Land Animal
The oldest known terrestrial animal fossil from New Zealand is of Middle Triassic age. The fossil is a single bone found near Mataura Island, due east of Invercargill, and has been identified as a shoulder bone from a large salamander-like amphibian, an ancestor of the frogs. Anne Warren of Monash University in Melbourne, an authority on Early Mesozoic Era amphibians, identified the fossil bone. She determined that it is from a labyrinthodont, a particularly common and diverse group of amphibians known globally from Permian and Triassic rocks of Gondwanaland and Laurasia. In this case, the actual animal would have been several metres in length. This exciting find was made in the mid-1990s by Hamish Campbell.

Fossil bones of marine reptiles have been collected from Middle Triassic rocks in the Wairaki Hills (Murihiku terrane) and Mt Harper (Rakaia terrane). An enigmatic fossil reptile limb bone is also known from Stephens Island (Dun Mountain–Maitai terrane) and is likely to be of Middle Triassic age.

Dicroidium Flora
Fossil plants of Middle Triassic age are of particular interest because they are dominated by a group of plants that is utterly different from the distinctive Permian *Glossopteris* flora of Gondwanaland. This is the *Dicroidium* flora that is well known from eastern Australian rocks. It forms another strong link with Gondwanaland. In New Zealand, the *Dicroidium* flora has been well documented by Greg Retallack. Fine specimens have been collected from several localities in the Wairaki Hills, Southland (Murihiku terrane) and near Benmore Dam in north Otago (Rakaia terrane). The youngest rocks in the Dun Mountain–Maitai terrane are of Middle Triassic age.

Kaka Point
Rocks of Middle Triassic age are exposed in south Otago along the Kaka Point to Nugget Point coast. This is the only place in the South Island where there are coastal exposures of rocks of this age.

The rocky shore near Kaka Point is spectacular. The wave-cut shore platform resembles a pyjama-stripe cloth with thousands of layers of grey sandstone, darker siltstone and pale coloured layers

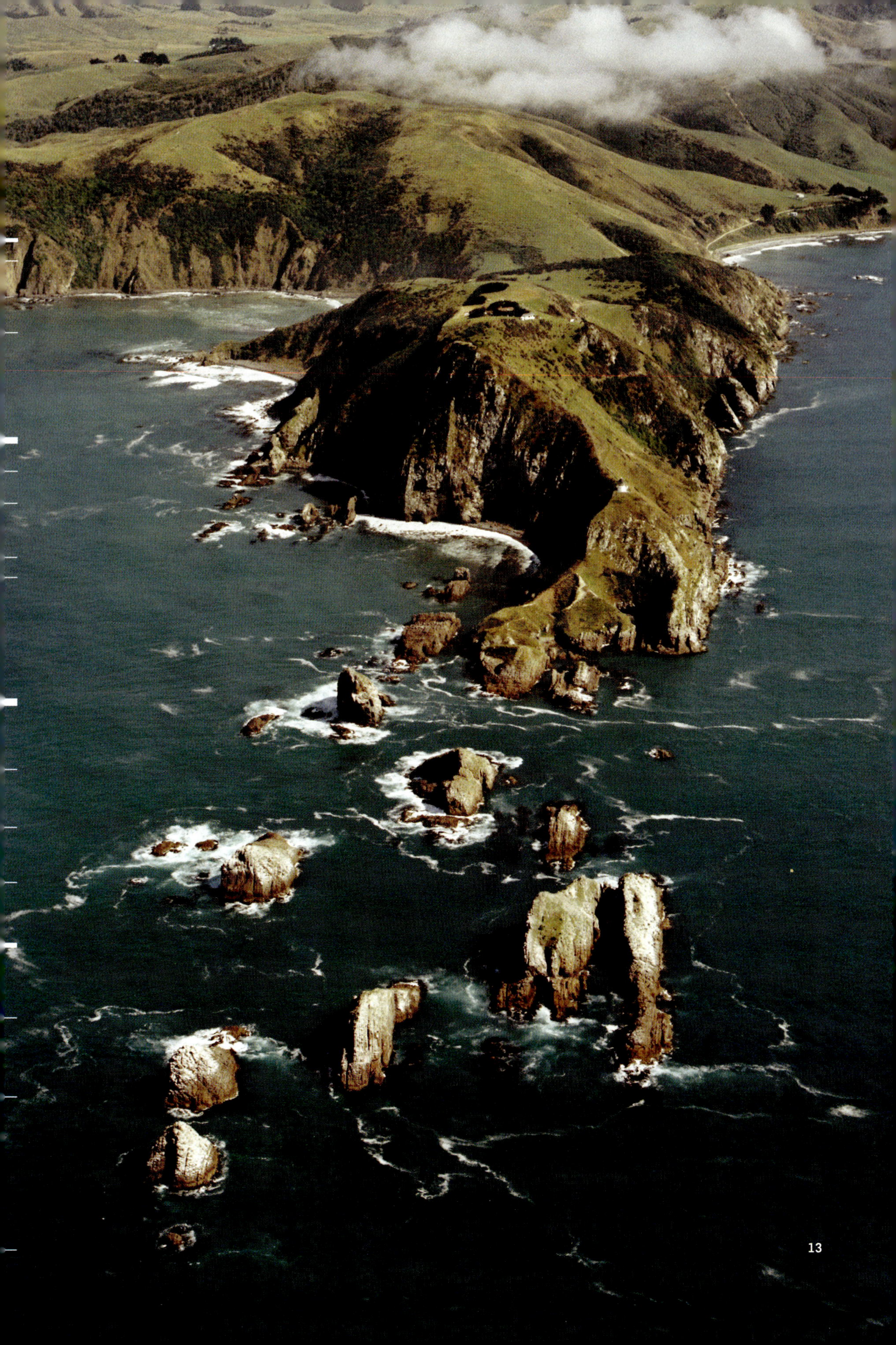

13. The Nuggets and Nugget Point, on the Catlins coast, south of Kaka Point, looking to the west. Although best known for its lighthouse, yellow-eyed penguins and seals, it is just as famous for Roaring Bay, on the south side of the point. Here, marine fossils of Late Triassic age are well exposed. The avid fossil collector is bound to find something of interest.

14. Jack Grant-Mackie (University of Auckland), one of New Zealand's most influential and best known palaeontologists with wide interests in teaching, Mesozoic molluscs and fossil birds.

15. A widespread and distinctive clam that existed almost globally during a portion of Late Triassic time: *Monotis*. Meaning 'single ear', this shell is ear-shaped and also sports a small ear or flange of shell along the hinge-line where the two valves of the shell are attached. This ear protects the attachment organ, the byssus. First recognised from New Zealand by Ferdinand von Hochstetter in 1859, *Monotis* has been studied by Jack Grant-Mackie. It is best-known from shellbeds on the west coast south of Auckland near Kiritehere, but occurs in many places elsewhere. Significantly, it is the symbol of the Geological Society of New Zealand, established in 1955.

of volcanic ash (now tuff). The rocks are beautifully stratified but have been tipped on end so that all the beds are now vertical. Despite having been deformed by folding and faulting, these rocks have not been subjected to high-grade burial and metamorphism. Fossils are not especially common, but nevertheless they are there. The most common fossils are the beaks of ammonoids. These are very similar to the beak of the modern octopus. Ammonoids are sought after by collectors but are not especially common. Other fossils include plant remains with *Dicroidium*, trace fossils, bivalves, gastropods, brachiopods, echinoids and crinoids. Microfossils have been recovered, including radiolarians and palynomorphs, but so far only one conodont has been found!

Late Triassic

Rocks of Late Triassic age (237–200 million years ago) occur in the Murihiku, Caples, Rakaia and Waipapa terranes. The youngest fossils known from the Caples and Rakaia terranes are of this age. Radiolarian fossils of latest Triassic age are known from Rakaia terrane rocks at the Orongorongo River mouth, east of Wellington. A youngest Triassic age of sediments in the Caples terrane is based on detrital zircon population studies. In other words, the age determination of about 100 sand grains of the mineral zircon from any one sample of Caples terrane greywacke, shows that there are zircons of Triassic age but no younger. And because it is impossible for sediment to be younger than the rocks it is derived from, we can be certain that the greywacke in this case is not older than Triassic. Other geological evidence allows us to be certain that it is not younger than Jurassic. Zircons that are about 202 million years old, of latest Triassic age, are present in Caples terrane sediments, therefore they must have accumulated as sediment about 200 million years ago. The age of sediment accumulation must be younger than the zircon age. The zircons grew as crystals within granite and the granite had to then be eroded to produce zircon-bearing sediment.

Fossil-bearing rocks of Late Triassic age are widespread in New Zealand, especially Southland, Otago, Canterbury and Nelson in the South Island. Yet, there are only a few places where rocks of this age are exposed on the coast and are therefore easily accessible and visible to the public. In the South Island, the only place is in Roaring Bay, south of Nugget Point along the south Otago coast. In the North Island, the only place is on the so-called King Country coast between Awakino and Albatross Point. Rocks exposed on the south Wellington coast of Cook Strait are also of Late Triassic age, but fossils are extremely rare.

Late Triassic Fossils

A series of very extensive Late Triassic shellbeds can be traced throughout Southland. They form a distinctive sequence that has been well studied, mapped and documented over the past 150 years. These fossil shellbeds are dominated by the shells of extinct clams and mussels.

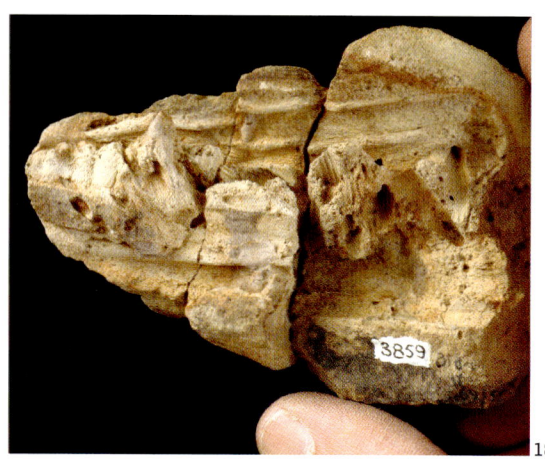

They have wonderful names such as *Oretia*, *Halobia*, *Manticula*, *Monotis* and *Otapiria*. To form shellbeds, often metres thick and extending over hundreds of kilometres, the shellfish must have existed in vast numbers and are therefore referred to as 'gregarious' bivalves. They have also been referred to as 'opportunistic', because they have seemingly taken the opportunity to smother the sea floor almost to the exclusion of other species. All that is certain is that they preserved well! There are in fact many other fossils present, including brachiopods, gastropods, scaphopods, ammonoids, nautiloids, bryozoans, echinoids, crinoids, conulariids, foraminifera, crustaceans such as crabs and ostracods, radiolarians, plant fossils and trace fossils.

Spores and Pollen

Fossil spores and pollen are well preserved in Triassic sediments of the Murihiku terrane and these have been well studied by Ian Raine (GNS Science) and Noel de Jersey (formerly of the Geological Survey of Queensland, Brisbane, Australia). They have established the remarkable similarity between the fossil palynoflora of New Zealand and eastern Australia, helping confirm the close Gondwanaland relationship of the two types of flora. It seems they were one and the same.

Oldest Dinoflagellates

Among the Late Triassic palynoflora is the first appearance of an important group of microfossils belonging to marine plants (algae). These are the dinoflagellates. They are unusual in that they have organic walls made of a tough, cellulose-like compound rather than shell material. They may be regarded as the resting cysts of algae, representing just part of their life cycle. They are surprisingly common and easily extracted from rock using acids. Notably, Graeme Wilson (GNS Science), New Zealand's foremost authority on dinoflagellates, has described *Sverdrupiella* from latest Triassic rocks exposed in Roaring Bay, on the south Otago coast. These are about 203 million years old.

Ichthyosaurs

Fossil remains of fish and marine reptiles such as ichthyosaur have also been recorded from rocks of Late Triassic age, but they are rare. A famous fragment of jaw, complete with teeth, was found by Doug Campbell in the early 1960s from rocks in the Otamita Stream, Hokonui Hills, Southland. This is the most spectacular Triassic ichthyosaur find known from New Zealand, but confirmed identifications have been made on fossil bones from a number of localities at Mt Potts in Canterbury

16. *Rastelligera elongata*: a Late Triassic brachiopod with a distinctive 'rastellum' or comb-tooth structure along the hinge-line. No other brachiopod has this bizarre interlocking structure. This specimen is an internal mould, that is, a natural impression of the inner surface of one valve of the shell. It is preserved in sandstone and was collected from rocks in the Taringatura Hills, Southland.

17. Ian Raine, palynologist with a particular interest in the Mesozoic pollen floras of New Zealand, the floral record of our coals. He is one of the masterminds behind the online electronic New Zealand Fossil Record File, a database listing all known fossil localities in New Zealand.

18. Ichthyosaur. Found in the late 1960s by Doug Campbell. This is a rare find. It is an external mould of a fragment of jaw, complete with a few teeth. It is preserved in siltstone of Late Triassic age from the Otamita Valley, Southland. The teeth are characteristic of ichthyosaur, marine reptiles that dominated Mesozoic seas.

19. Albatross Point, to the west of Kawhia Harbour, looking south towards Taharoa, Marokopa and Kiritehere. Part of the Kawhia Syncline, the rocks of this coast offer some of the finest exposures of Triassic–Jurassic Murihiku Supergroup rocks in the North Island.

(Rakaia terrane) and the Hokonui Hills in Southland (Murihiku terrane)

Taranaki and Wellington Coasts

In the North Island, Late Triassic rocks of the Murihiku terrane are even better exposed and over a much greater length between Marokopa and Tirua Point on the north Taranaki coast of the Tasman Sea. This coast is not especially accessible, however. The best-known and most accessible localities are along the Marokopa to Kiritehere coast.

The Wellington coast of Cook Strait is also of Late Triassic age, as is Kapiti Island, but here the rocks are all greywacke within Rakaia terrane. Fossils are rare and inconspicuous. You have to know what you are looking for, and no fossil shellbeds are known. The most common fossils seen are tube fossils of *Torlessia*, which are thought to be foraminifera. The age of these rocks is well established on the basis of radiolarian microfossils extracted from phosphatic nodules. As you go from west to east along the south coast of the North Island, the rocks become progressively younger, and the youngest Triassic fossils are from outcrops near the bridge over the Orongorongo River.

Fossil plant remains are present in the greywacke around Wellington, but invariably are not able to be identified, as the preservation is too poor. No fossil palynomorphs are preserved either because the rocks are too metamorphosed.

Northernmost Triassic

Late Triassic *Halobia* limestone has been recorded from the northern coast of Stephenson Island and adjacent Cone Island in Whangaroa Bay, Northland. These rocks are within Waipapa terrane and they constitute the northernmost old basement rocks of New Zealand. There are no known rocks on land to the north of Triassic age or older: they are all younger.

Murihiku

By far the best place to look for Late Triassic fossils is in Murihiku terrane. A belt of Late Triassic rocks runs from the south Otago coast inland towards Gore and through the Hokonui, Taringatura and Wairaki Hills of Southland. It is interesting to note that the highest topographic point within Triassic rocks of Murihiku terrane is the top of Mt Hamilton at the northern end of the Takitimu Mountains.

Alpine Triassic

The highest point in New Zealand, Aoraki/Mt Cook (Rakaia terrane), is almost certainly of Late Triassic age. Large tracts of the greywacke in Canterbury are of Late Triassic age, with numerous fossil localities of mainly *Monotis* or *Torlessia* known from the Arthur's Pass area in particular, but also elsewhere. An easy place to find *Torlessia* is on the Mt Hutt skifield road.

A spectacular Late Triassic fossil shellbed dominated by large shells of trigoniid is known from the Mathias River, a major tributary of the Rakaia River, almost on the main divide right in the heart of the Southern Alps. Trigoniids are markedly triangular in shape, have thick shells and a powerful foot. They were rapid burrowers, not unlike some modern-day surf clams. The trigoniiids are all but extinct, with only one or two species known from Tasmania today (*Neotrigonia*). At least two different trigoniids are present in the Mathais: *Caledogonia*, which is smooth, and *Maoritrigonia*, which is ornamented with radial rows of tubercles or raised knob-like structures.

Triassic Ends with Mass Extinctions

As with the Permian–Triassic and younger Cretaceous–Paleogene boundary, there is great interest in the Triassic-Jurassic boundary (the Jurassic lasted 200–146 million years ago). Many organisms became extinct at this boundary, more so than at the Cretaceous–Paleogene boundary 65 million years ago. There are a few places where you can see this boundary in New Zealand. Firstly, it is present in places in Southland such as Taylors Stream in the Hokonui Hills; secondly, the south Otago coast, south of Roaring Bay; and thirdly, it is in rocks exposed in the Awakino Gorge and along the Marokopa–Kiritehere coast of the western North Island. These three areas are within Murihiku terrane, and the location of the boundary has been determined on the basis of shelly fossils, fossil pollen and spores, and also carbon isotope signatures derived

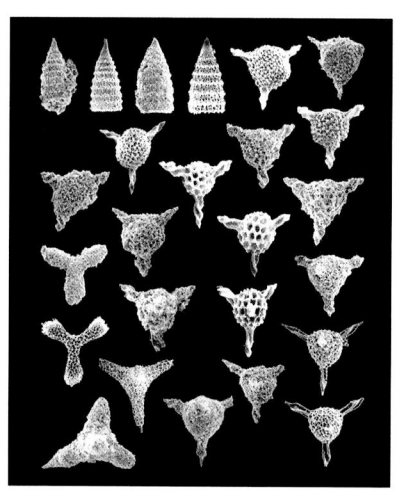

20

from organic (plant) material.

Lastly, there is a locality on Pakihi Island in Tamaki Strait, Hauraki Gulf, near Auckland. These rocks are very different from the Murihiku rocks that would have accumulated in water depths of less than 200 metres. They are deep ocean sediments of the Waipapa terrane and the position of the boundary is based on radiolarian fossils.

By the end of Triassic time, many ancient groups of organisms became extinct, including the conodonts and conulariids. All but a handful of brachiopods and ammonoids survived whatever crisis was responsible for the boundary. The ammonoids that did survive all had ammonitic suture. The suture pattern of ammonoids has been broadly described in terms of three principle styles, depending on how frilly they are, starting with relatively simple goniatitic suture (named after the ammonoid genus Goniatites) in Ordovician to Permian ammonoids. More complex ceratitic suture (named after the genus Ceratites) developed in ammonoids of Permian–Triassic age, and highly ornate ammonitic sutures (named after genus Ammonites) developed in Late Triassic, Jurassic and Cretaceous ammonoids.

Ammonites

Ammonites are a subgroup of ammonoids that have a distinct suture pattern referred to as 'ammonitic'. They first appeared in Late Triassic time. Triassic ammonoids were dominated by forms with ceratitic suture, and prior to the Triassic in the Permian and Carboniferous, ammonoids had goniatitic sutures. Note that the 'suture pattern' is the distinctive pattern formed at the intersection between the tubular coiled spiral shell of an ammonoid and the walls of the chambers, referred to as 'septae', that grow at regular intervals as the shell grows but at right angles to it. The goniatitic ammonoids are regarded as the more primitive and the ammonitic as advanced, the latter surviving right through to 65 million years ago, becoming extinct with the dinosaurs.

The suture pattern can be incredibly beautiful. As is characteristic of the intersection between any two surfaces, it forms a line: thin, sinuous and highly ornate. It appears as a complicated pattern. Ammonite paleontologists have developed an equally elaborate language to describe suture patterns. The reason for the extraordinary complexity of the suture pattern is that it represents a huge surface area of attachment between the wall (septum) and the shell, thereby adding enormous strength. Architects and building inspectors can but be inspired by this wonder of nature!

The Jurassic: Ancestral New Zealand Takes Shape

The Jurassic Period (200–146 million years ago) ushered in profound changes, as the New Zealand region of Gondwanaland gradually shifted from near the South Pole into more temperate latitudes. Largely submerged up until now, this area rose as a result of massive earth movements, known as the Rangitata Orogeny (mountain building). By the Middle to Late Jurassic Period, an ancestral New Zealand had emerged as a broad margin skirting the eastern side of Gondwanaland, almost half the size of present-day Australia.

Now the stage was set for the spread of Gondwanan plants and animals across this new land mass. These would have included the ancestors of the distinctive, modern-day New Zealand flora and fauna: the podocarps (precursors of today's rimu, totara, miro and kahikatea), araucarian pines (precursors of modern kauri), amphibians (precursors of the native frog *Leiopelma*), reptiles (including ancestors of the tuatara), the worm caterpillars (like *Peripatus*), worms and insects (ancestors of the weta).

20. Exquisitely preserved, tiny skeletons of representative fossil radiolarians of Late Triassic age. Scale bar is 0.1 mm in length. These fossils were extracted from phosphorite within greywacke exposed on the Wellington coast of Cook Strait.

21. Rolling sheep country of the Hokonui Hills, Southland, looking east from near Ram Hill between the Oreti and Mataura river valleys. The underlying rock in this view is Triassic Murihiku Supergroup and it is part of the steeply inclined north limb of the Southland Syncline. The sandstone formations appear more prominent, accentuated by shrubs (matagauri), whereas the siltstones are clothed in pasture or tussock.

22. *Lytoceras taharoaense*: a Late Jurassic ammonite from Kawhia Harbour, south Auckland, with its distinctive ammonitic suture pattern highlighted in paint. This complex, highly ornate pattern is formed at the intersection of the septum (chamber wall) with the whorl interior (inside the shell tube). The white paint occupies space between one septum (black line) and the previous septum. The organ that secretes the calcite septum and shell is referred to as 'the mantle' and must rate as one of nature's marvels.

Birds also appeared. During the Jurassic Period the ancestors of the large, flightless ratites evolved in Gondwanaland. Nowadays, their descendents are in Africa (ostrich), South America (rhea), and Australia (emu and cassowary). New Zealand's ratite representatives are the kiwi and the now-extinct moa. However, we must not forget that there is very little fossil evidence of these ancestral birds, and with the passage of time, they may have changed dramatically in appearance and size. It is unlikely that moas, for instance, looked like moas in Jurassic time. They were probably very different.

Jurassic Rocks, Molluscs and Brachiopods

Jurassic rocks are moderately widespread in New Zealand but not nearly as much as Triassic rocks. They are best known from Murihiku terrane rocks exposed in Southland and the Kawhia Harbour area on the south-west Auckland coast of the Tasman Sea. These rocks are especially fossiliferous and have been intensively studied by numerous geologists and paleontologists over the past 150 years.

Jurassic rocks occur in the following Eastern Province terranes: Brook Street, Murihiku, Waipapa and Pahau. Middle Jurassic rocks are the youngest rocks in the Brook Street and Waipapa terranes, whereas Late Jurassic rocks are the youngest in the Murihiku terrane.

These Jurassic rock sequences are characterised by a series of molluscan shellbeds dominated by bivalves with names such as *Pseudaucella*, *Meleagrinella*, *Malayomaorica*, *Retroceramus* and *Australobuchia*. Associated with them are cephalopods, notably ammonites and belemnites. Less common are the brachiopods and a variety of other shelly fossils.

Jurassic brachiopods, in particular the rhynchonellids, have been studied by Donald MacFarlan (formerly of the University of Otago). This group is especially diverse and well represented in the Jurassic Period. Although the brachiopods survived the Triassic–Jurassic boundary, some became extinct, in particular the athyrids, which were especially common in the Late Triassic Period. The spiriferinids managed to survive until the end of Early Jurassic time (200–176 million years ago) and then they were gone forever also. All that remained were the terebratulids and the rhynchonellids, and these are the groups that still exist today.

Belemnites

Globally, belemnites first appear in Late Triassic time, but they are first recognised in New Zealand in Early Jurassic rocks. These belemnites have been well studied by Graeme Stevens (GNS Science) and Brian Challinor. Numerous belemnite species are known from Jurassic rocks, and along with bivalves and ammonites, they permit a well-established subdivision of our Jurassic rocks, just as they do elsewhere.

Belemnites are bullet-shaped, bullet-sized rods of calcite with distinctive radiating crystal structure. They preserve well in rock, and are surprisingly common in places. They may be thought of as hard parts or skeletal structures that permitted the attachment of muscles deep within the animal. However, it is thought that they evolved within squid-like cephalopods as part of an on-board, weight-belt-like strategy for rapid descent within the ocean. No doubt they could counter the weight-belt effect by inflating suitable body cavities with gas for rapid ascent.

Jurassic Ammonites

The Jurassic Period is well known for its ammonites. In New Zealand they have been studied extensively by many paleontologists, including Graeme Stevens, Gerd Westermann, Neville Hudson and Jack Grant-Mackie. The ammonites permit remarkably precise age determination. For instance, the Early Jurassic ammonite *Psiloceras* ranges in age from 199.5–199.0 million years ago. We know the age range from detailed studies of Jurassic rocks and the succession of ammonites globally, especially in Europe, but also elsewhere, such as in Argentina.

Jurassic shelly fossils include brachiopods, molluscs, bryozoans, echinoids, crinoids, corals, crustaceans such as lobsters and ostracods, foraminifera,

23. *Dimitobelus lindsayi*: Belemnites preserved in sandstone from Late Cretaceous rocks near Haumuri Bluff in north Canterbury. This rock has been cut with a diamond saw to better show the orientation of the fossils in cross section. Belemnites are bullet-shaped, solid tapering rods of calcite that were produced by a group of squid that existed for much of Mesozoic time, only to become extinct at the end of Cretaceous time along with the dinosaurs and marine reptiles. Belemnites served as built-in weight belts for their owners.

24. *Mecochirus marwicki*, a rare find. This fossil lobster was collected near Kawhia Harbour from mudstone of Late Jurassic age by Ron Gardner. It is about 155 million years old.

25. Graeme Stevens, a world authority on Jurassic belemnites and ammonites and the man who brought New Zealand geology to the New Zealand public through his books and articles.

26. The giant ammonite. This is an exceptionally large specimen of *Lytoceras taharoaense*, the largest Jurassic ammonite ever collected. It was found by Jean Gyles in the 1970s near Taharoa, just south of Albatross Point, and was extracted from the rock using explosives.

27. Brendan Hayes discovered and described a single small dinosaur bone from Late Jurassic rocks near Port Waikato. This is the only Jurassic dinosaur recorded from New Zealand so far. The fossil is a finger or toe bone of a small theropod.

The Younger Old Rocks: Eastern Province

28. *Notohaglia mauii*, a fossil wing case of a weta-like insect. This is the only Mesozoic insect fossil known from New Zealand. This remarkable specimen is from Late Jurassic rocks (Murihiku terrane) exposed near Port Waikato on the western coast of the North Island.

29. A fallen giant on a Middle Jurassic fossil forest floor, here exposed on the shore platform in Curio Bay on the Catlins coast, Southland. A number of logs and tree stumps are preserved here as silicified fossils, possible casualties of a Mt St Helens-style volcanic eruption.

radiolarians, calcareous nannofossils and calcispheres. Vertebrate fossils are virtually unknown from Jurassic rocks of New Zealand. No marine reptile bone fossils have been recorded.

A Single Jurassic Dinosaur

One small theropod dinosaur bone, a finger bone, is known from Late Jurassic rocks near Port Waikato. This was discovered by Brendan Hayes, a keen fossil hunter. One other vertebrate fossil is known from Jurassic rocks in New Zealand. This is an enigmatic reptile bone of Middle Jurassic age that was collected from fossil plant beds near Nelson.

The Late Jurassic was the time of the largest known dinosaurs, some exceeding 40 metres in length. It was at this time that the first true birds appeared: feathered and beaked theropods capable of flight. However, fossils of such animals are not known from New Zealand, and, but for a few scales, there are virtually no fish fossils either.

Who Left that Heap?

Some strange, lumpy fossils have been found from the Huriwai Beds at Port Waikato. Although enigmatic, it is just possible that they are coprolites, the fossil remains of animal droppings. Given their size, however, they would have to relate to a large animal, possibly and not improbably, a dinosaur.

Jurassic Weta

Almost no insect fossils are known, but there is at least one. A single, left wing of an ancestral locust, and possibly ancestral to the wetas as well, named *Notohaglia mauii*, has been described from Late Jurassic rocks near Port Waikato.

Palynomorphs and Paleobotany

Jurassic palynomorphs have been studied by Ian Raine and Noel de Jersey in particular. There has been more detailed study of fossil leaf florae and also fossil wood, which are well represented in Middle and Late Jurassic rocks. Fossil forest floors, complete with paleosols (fossil soils) and fallen logs, are preserved in numerous places within the Catlins area of Southland and also the Port Waikato area in the North Island.

Curio Bay

Along the Catlins coast of Southland, there are extensive rock exposures of Middle Jurassic age (176–157 million years ago). Most famous is the Curio Bay fossil forest on the Catlins coast. This is a classic example of a forest being preserved by rapid burial, in this case, probably an eruption, with subsequent flooding and rapid burial by sediment. What is actually preserved is a fossil forest floor, complete with tree stumps, fallen logs, leaf litter and soil.

The fossil logs have been silicified; in other words the wood has been replaced by silica. This happens when wood is rapidly buried in sandy sediments. Percolating groundwater enriched in silica preferentially 'flows' through wood as if the logs were giant straws, and the silica precipitates within spaces once occupied by organic tissues of the wood. In the 1990s, Vanessa Thorn made detailed studies of the Curio Bay fossil flora. The logs are those of conifers, but what is most interesting is that they grew at a time when Southland was very near the South Pole, so the forests of the day would have experienced darkness for a good part of the year. The fossil logs at Curio Bay must rate as perhaps the largest known fossils from New Zealand.

Pahau and Brook Street Terranes

Middle to Late Jurassic sediments of the Pahau terrane occur in the catchment of the Ashburton River. These rocks are referred to as the Clent Hills Group. They include leaf beds and thin coal deposits associated with river sediments and marginal marine sediments. Of particular interest as they might be prospective for dinosaur hunting, they appear to represent rocks that may be much more widespread along the eastern side of New Zealand but are largely buried by younger sediments.

The Jurassic record in the Brook Street terrane is meagre, but nevertheless significant. It includes shallow marine formations in the Wairaki Hills, Southland (Barretts Formation), in the Barrier range in northern Fiordland (Barrier Formation), and probably in the Nelson area (Drumduan Group).

Limestones (Pahau)

Rare Late Jurassic limestones that accumulated in open oceanic settings are known from at least four localities in Pahau terrane rocks. The first locality is near Kaiwara, west of Cheviot in north Canterbury. The other three localities are in the North Island. One is on the Wairarapa coast at Mukamuka; another is inland at Ekatahuna; and a third is in the Ikawhenua Range near Whakatane. These are most interesting limestones, with rich and diverse fossil faunas, including bivalves, gastropods, brachiopods, corals, crinoids, foraminifera, dinoflagellates, radiolarians, calcispheres and coccoliths. Curiously, very few ammonites have been found in these rocks. Nevertheless, these rare discoveries convey a huge amount of information about marine life in the ocean at the time. The fossils that make up the limestone are quite different from the Middle to Late Jurassic fossils of the Murihiku terrane. They have been studied by Phillip Maxwell in particular.

The Pahau terrane limestones relate more to the Panthalassa Ocean than the margin of Gondwanaland, and are almost certainly best interpreted as remnants of guyots, flat-topped

30

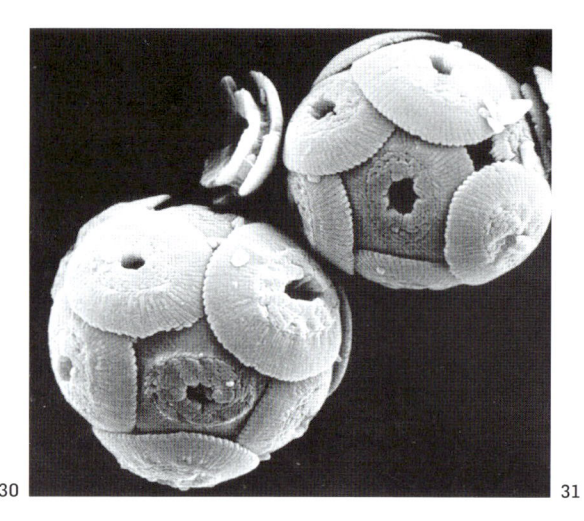

31

underwater mountains that originated as volcanoes (they may be thought of as 'spent cones', rather like spent fireworks). The limestone is invariably associated with basaltic pillow lavas and volcanic sediments that erupted and accumulated in an entirely submarine environment. Sea-floor spreading and subduction processes have tectonically off-scraped and juxtaposed these long-distance travellers within the greywacke sediments of the Gondwanaland margin.

Pahau terrane greywackes are not especially fossil-rich, but there are some localities in north Canterbury, Marlborough and the central North Island ranges and the Hauraki Gulf.

Final Assembly

The old Cambrian to Early Cretaceous basement rocks of New Zealand had assembled into their present configuration by no later than 130 million years ago. We know this because they are overlain by younger sediments that are not older than 130 million years.

The Cretaceous: A Momentous Geological Period

Many New Zealand geologists say that the key to understanding New Zealand geology is our Cretaceous geology. The Cretaceous Period (146–65 million years ago) was momentous for New Zealand. After all, it is in the Early Cretaceous age that the Zealandian sector of the eastern margin of Gondwanaland was assembled and bulldozed into place to become a continental land area, albeit just part of the Gondwanaland land mass.

All was well until the stirrings of continental rifting, stretching and thinning began about 125 million years ago. Then, in the Late Cretaceous, 83 million years ago, Zealandia achieved clean separation from Gondwanaland with the formation of the Tasman Sea floor.

Early Cretaceous Marine Fossils

The Eastern Province includes the Early Cretaceous Pahau terrane, which in turn is overlain by younger Cretaceous marine sediments. These are fossil-bearing and are dominated by molluscs, in particular the inoceramid clams. Harold Wellman (formerly of The New Zealand Geological Survey) was one of the first to study these fossils and it was he who first worked out how many different species are represented. He also worked out the order in which they occur through geological time. This research has subsequently been championed by James Crampton at GNS Science.

Early Cretaceous marine fossils that relate to Gondwanaland and the Panthalassa Ocean include not just clams but also ammonites, belemnites, gastropods and rare brachiopods. A few rare bone fragments are known and are almost certainly all of marine reptile origin. Microfossils of Early Cretaceous age include dinoflagellates, pollen and spores.

Marine rocks of Early Cretaceous age are not especially widespread, and nor are the fossils contained within them especially varied or abundant. They can be found in a few places in Southland, the West Coast, Otago, north Canterbury, Marlborough, the Wairarapa, the Raukumara Peninsula and Northland. The best exposures are in remote parts of Marlborough and the Urewera Ranges of the Raukumara Peninsula.

These rocks are important because they offer the best record of marine life along the east coast of Gondwanaland during the 35-million-year interval prior to separation of Zealandia from Gondwanaland. At this time in the history of Gondwanaland, the Zealandian sector was located in cool southern latitudes, straddling the Polar Circle at 55–70°, and although there were no polar ice caps at the time, the climate was probably not all that conducive for good living.

32

Tapuae-o-Uenuku and Mt Somers

The modern New Zealand South Island landscape sports the remnants of at least two mid-Cretaceous volcanoes: Mt Somers on the western margin of the central Canterbury Plains; and Tapuae-o-Uenuku, at 2885 metres the highest peak in the Inland Kaikoura Mountains, and the highest peak in New Zealand north of Arthur's Pass. Tapuae-o-Uenuku is clearly visible from the Wellington coast of Cook Strait.

These are all that remain of subduction-related volcanoes that erupted 100–90 million years ago. At the time, the Chatham Rise was part of the eastern margin of Gondwanaland and it was not unlike modern South America: a great elongate margin between continental crust on the Gondwanaland plate and oceanic crust on the Panthalassa Plate or Paleo-Pacific Plate. The Chatham Rise would have supported subduction-related volcanoes all along it. Mt Somers and Tapuae-o-Uenuku may be regarded as inland or westernmost volcanoes within a segment of the mid-Cretaceous arc, or Ring of Fire, bounding Gondwanaland.

Mt Somers has a volcanic form to it, but this is a fortuitous artefact of erosion. It is comparable to Tauhara, which forms the upstanding but small volcanic mountain to the immediate west of Taupo. The two mountains are even of similar dacite composition. Dacite is the term for volcanic rocks that have a higher silica content than andesite but a lower silicon content than rhyolite. However, Tauhara erupted only 16,000 years ago, and Mt Somers is the remnant of something much bigger that erupted 90 million years ago.

Tapuae-o-Uenuku is actually greywacke, but it owes its massive girth and upstanding height to the roots of a volcano. The greywacke is riddled with igneous dikes that bolster the greywacke and add strength, rather like steel reinforcing does to concrete. The dikes run across the landscape like walls, just as the word 'dike' suggests, but are no greater than a few metres wide. These are the trackways of lava that have flowed upwards from a deeper-seated magma chamber, cutting their way to the surface along fissures that probably opened up because of inflation of the crust from the growing volcano below. The volcano has been eroded away, but the feeder roots or conduits survive as dikes.

To gaze upon Mt Somers or Tapuae-o-Uenuku is to stare at the magnificent stumps of two volcanoes that relate to our Gondwanaland heritage. They erupted and became extinct prior to the separation of Zealandia.

The following chapters (in Part 3) will follow the voyage Zealandia took as it drifted away from Gondwanaland and then describe the special cargo of plants and animals that floated aboard it.

30. Nanofossils. These tiny spheres of intricate calcite plates are produced by single-celled plants (algae). Chalk is comprised of nanofossils.

31. *Cremnoceramus bicorrugatus*, a substantial meaty clam that dominated Cretaceous seas. This specimen was collected from rocks near Coverham, north Canterbury. It is 250 mm long.

32. A cobble found on a beach in Waihere Bay, Pitt Island, in the Chatham Islands. This is a piece of silicified fossil wood of Cretaceous age, about 100 million years old. The grain of the wood is superbly preserved and closely resembles modern pine! The cobble is 160 mm long and 100 mm high.

Part 03:

Zealandia:
83–23 MILLION YEARS AGO

09/ Zealandian Dinosaurs

A Full Complement Of Cargo

Questions flood to mind. When Zealandia broke free of Gondwanaland, what did it look like? What shape did it have? What was the landscape like? What life was there? The last question is easy to answer: Zealandia would have supported a full complement of plants and animals that were characteristic of eastern Gondwanaland at the time, about 83 million years ago. The animal world would have been dominated by dinosaurs, flying reptiles, birds, mammals and amphibians. The plant world was dominated by conifers and ferns, but there were also flowering plants. The grasses and many other plants had yet to evolve.

Once the two continents were separated by sea, like two ships parting company, direct

1. Around 83 million years ago, Zealandia, the land that ultimately became New Zealand, parted company with Gondwanaland, taking with it plants and animals identical to those of its giant host continent. For 20 million years the spreading sea floor ridge to the west continued to widen to create the Tasman Sea; about 63 million years ago this movement ceased, leaving Zealandia isolated in the South Pacific.

communication between the two live cargoes would have all but ceased. Relatively easy contact would have been maintained by some of the flying reptiles and birds. In the sea, the same would have been true of the long-ranging pelagic (that is, swimming or floating) molluscs (ammonites and nautiloids), fishes, turtles and marine lizards. Then, of course, there would have been those organisms that chance fate, willingly or unwillingly, and drift afloat on the sea or in the wind.

Like two giant biological clocks, both cargoes would have evolved in their own way. Life on Gondwanaland would have been different from life on Zealandia, assuming that they were truly distinct and separated by deep sea.

It is easy to imagine Zealandia, almost half the size of Australia, with its plants and animals just like those of eastern Gondwanaland. They would have been identical to the Gondwanan fauna and flora, at least initially. Over time, however, changes would inevitably have developed, and animals and plants distinct from those on Gondwanaland would eventuate. The random walk of evolution would have kicked in. We can therefore be certain that there must have been a distinctive fauna and flora (collectively known as a 'biota') that was peculiar to Zealandia.

By the time Zealandia departed 83 million years ago, Gondwanaland was much diminished, but nevertheless still comprised Australia and Antarctica. India had departed about 100 million years ago. South America and Africa had headed off together about 160 million years ago, and subsequently parted company about 130 million years ago. Australia was the last to leave, separating from Antarctica much later, at about 40 million years ago. We know all this from detailed study of the age distribution of the basalt forming the ocean floor, which is largely based on the magnetic signature preserved in the rocks.

Our Excuses for Ignorance

Curiously, the concept of a distinct post-Gondwanaland but pre-New Zealand 'Zealandian biota' has not really been entertained or discussed much, particularly within the mainstream science world. This concept is relatively new. There are several reasons for this.

Firstly, it has always been understood that there must have been continuous land present in the New Zealand region since Zealandia broke away

2. *Moanasaurus mangahouangae.* Measuring 680 mm in length, this skull was that of a mosasaur, the dominant sea predator of its time. Mosasaurs were marine reptiles, not dinosaurs, with large skulls and short necks, and their fossil bones have been discovered in Cretaceous rocks in Hawke's Bay, Marlborough, Canterbury and Otago.

3. *Tuarangisaurus keyesi.* The 370-mm-long skull of a long-necked elasmosaur, a group of marine reptiles, not dinosaurs, that flourished in late Mesozoic seas globally. On display at Te Papa, this was collected by Joan Wiffen and her colleagues, from the same sandstone formation as the dinosaur fossils.

4. New Zealand's first three dinosaur fossils, now on display at Te Papa. All three were found by Joan Wiffen and her colleagues in the 1970s. A group of fossil hunters, they were deliberately searching for fossil bones in the Managahouanga Stream, a tributary of the Te Hoe River, inland Hawke's Bay. The very first bone to be recognised as dinosaur (in 1980) was the vertebra (top right). Australian vertebrate paleontologist Ralph Molnar considered it to be from the tail of a moderately large theropod.

from Gondwanaland. It has been almost unthinkable to suggest otherwise. How else can our rather strange modern native flora and fauna be explained? It has strong Gondwanan affinity and yet it has oddities that argue strongly for long isolation. Secondly, the fossil record of life in New Zealand for the past 83 million years, since it broke away from Gondwanaland, is largely a marine record. There is a terrestrial record, but it is much less obvious. Accordingly, it has not received as much research attention as the more conspicuous marine record. Thirdly, and lastly, the idea of 'Zealandia' as a substantial continental land area is relatively new. Only recently have we been able to appreciate just how large it is or was. Zealandia may have been an island, but it was of continental proportions, at least ten times the size of New Zealand today.

A Generation Gap
The significance of recognising a Zealandian biota, as opposed to a New Zealand biota, is that it permits the possibility of a more realistic and more complex history than previously thought. This is desirable, as the modern New Zealand biota demands better explanation in terms of the known geological history of New Zealand.

Irrespective of the geological constraints, there are obvious biological reasons for arguing a more complex history. Had there been continuous land over the past 83 million years, might we not expect to see much greater diversity in the modern New Zealand biota? For example, by contrast to the tropics or continental areas, New Zealand has a relatively small number of flowering plants and vertebrate animals. On the other hand, there is a high percentage of unique or endemic species. There are fewer of the so-called 'higher' plants and animals, but many species of spiders, snails, primitive insects, fungi, mosses and liverworts.

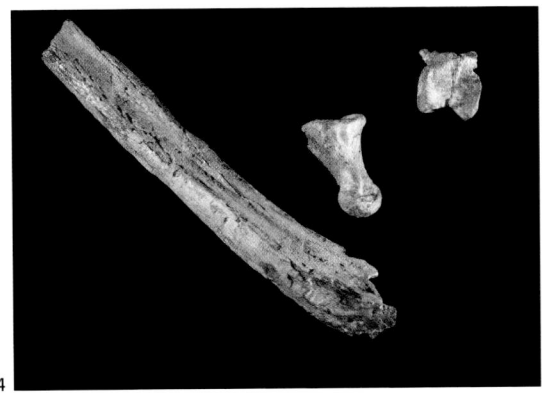

Our own species, *Homo sapiens*, has a remarkably short history, considerably less than 5 million years. Given what we now know about rates of evolution and speciation, imagine what might happen to a complex organism such as man in the space of 83 million years. The mind boggles.

It makes perfect sense to get the genealogy of our modern biota better sorted and understood. There are at least two steps that we can now be certain of: the New Zealand biota evolved from Zealandian biota; and Zealandian biota evolved from Gondwanan biota. Consider Zealandia the parent, and Gondwanaland as the grandparent.

To think that all these years we have assumed a distance gap of one generation from Gondwanaland, when all along it has been two generations!

The Zealandian Biota

What is this 'Zealandian biota'? What was it like? How is it recognised? Where is this biota now? Where can it be seen? What has happened to it? This unique ancestral living world, of continental proportions, all on its own, may explain some of the more unusual elements and peculiar oddities of the modern New Zealand fauna and flora. Or will it? We presume that the modern New Zealand biota has evolved directly from Zealandia, but can we be certain? And if it hasn't, what are the alternatives? Indeed, are there any?

These questions may seem odd, but our knowledge of the geological history of Zealandia clearly indicates that it was largely submerged by latest Oligocene time, 25 million years ago, and indeed geologically we cannot exclude the possibility that it was totally submerged. In which case, the Zealandian terrestrial biota might only have survived as refuges on small islands.

Such a scenario is hard to believe, and hard to prove, but it is fascinating to speculate about! What if . . . ? Let us consider further what we know of the Zealandian flora and fauna.

The Dinosaurs of Zealandia

In 1980 the world learnt of the first dinosaur fossil from New Zealand. This was major scientific news and announced at an international geological conference (called Gondwana 5) held in Wellington. Australian vertebrate paleontologists confirmed the identification of a small fossil bone that had been found by Joan Wiffen in 1975. Joan had discovered the fossil at a remote locality in a tributary stream valley of the Te Hoe River in inland Hawke's Bay.

This was an immensely important discovery. It meant that New Zealand must have had dinosaurs. Or did it? There was one small problem: the rock that the fossil was collected from is marine sandstone. It accumulated as sediment on the sea floor, and all dinosaurs were strictly terrestrial animals. Furthermore, there are plenty of fossil bones preserved in this particular sandstone, but they are mostly of marine lizards (ichthyosaurs, mosasaurs and plesiosaurs, which include

5. Trevor Worthy is a New Zealand vertebrate paleontologist with specialist interest in our Pleistocene fossil bird record of the last two million years. He is leader of the research team working on the St Bathans bone bed bonanza in central Otago and discovered New Zealand's first pre-Pleistocene fossil terrestrial mammal.

elasmosaurs and pliosaurs), turtles and flying lizards (pterosaurs). Perhaps the dinosaur bone had been part of a mosasaur meal, in which case the dinosaur bone could have been brought in from some other land mass.

Undeterred and more determined than ever, Joan Wiffen and her colleagues persisted with their labours and found what they were looking for: more fossil dinosaur bones. It can be said that Joan Wiffen's greatest scientific achievement is not so much that she discovered New Zealand's first dinosaur fossil, but that she discovered fossil evidence of a whole community of dinosaurs: she established without doubt that Zealandia had dinosaurs.

The implications of this realisation are profound. If there were dinosaurs, then there must have been other terrestrial animals, including reptiles other than dinosaurs, such as lizards and snakes, and most importantly, there must also have been mammals. As yet, no mammal fossils have been discovered at this locality, but they probably will be.

The reason for such confidence is simply this: dinosaurs and mammals have co-existed. They have lived alongside each other since their first appearance in Late Triassic time, about 225 million years ago. It is, therefore, entirely logical and reasonable to suggest that Zealandia must have had mammals.

As recently as 2006 it was announced that a primitive, mouse-like fossil mammal had been discovered in central Otago, dated at 19–16 million years. This is sensational news, because it is the first record of a pre-Pleistocene (that is, more than 1.8 million years old) terrestrial mammal from New Zealand – as opposed to a flying mammal such as a bat, or a marine mammal such as a seal.

The fossil locality was discovered some years ago by geologists Barry Douglas and Jon Lindqvist (from Dunedin), within lake sediments exposed in Falls Creek, near St Bathans in central Otago. The actual sediment is so richly fossiliferous in places that it is best described as a bone bed. Vertebrate paleontologist Trevor Worthy and colleagues have subsequently collected the site and have identified some of the bones, most of which are fish and aquatic birds such as ducks. Who knows what other exciting fossils may come to light!

Following from this thought, the next question is most interesting: why are there so few records of native, land-dwelling mammals in New Zealand today? Could they have been wiped out, or could the record have been destroyed? Surely it is very difficult to rid a continent, Zealandia in this case, of an entire group of animals such as mammals?

In today's world, mammals occupy every known major land mass, as well as most islands. They are immensely successful organisms, well adapted for terrestrial life in just about every conceivable habitat and environment, whether it is cold, wet, dry or hot. They are everywhere. But that is today. Perhaps things were very different in Zealandia, and, of course, the mammals were not nearly as evolved or as diverse as they are now, and nor were many other animal and plant groups.

The Hawke's Bay dinosaurs are of Late Cretaceous age, as is clearly evident from all the other fossils associated with the dinosaur bones, such as plants, pollen, plankton, shells and other bone fossils. But the Late Cretaceous age lasted a very long time, 35 million years, 100–65 million years ago. What we need is more age precision.

As luck would have it, a much more precise age is available. We can now be certain that the fossils are 75 million years old. This is largely based on fossil plankton, and in particular dinoflagellates. These are microscopic resting capsules produced by a particular group of marine algae. As luck would have it, the dinoflagellates have evolved rapidly, changing morphology and producing a succession of distinctive forms through geological time and especially during Late Cretaceous and early Cenozoic time.

This age of 75 million years for the Hawke's Bay dinosaurs is most interesting because it postdates separation of Zealandia from Gondwanaland by 10 million years. We can, therefore, be reasonably certain that the Hawke's Bay

Joan Wiffen – Dragon Lady

Dinosaurs once in New Zealand? Scientists did not seriously entertain the notion until several decades ago, when Hawke's Bay housewife and amateur paleontologist Joan Wiffen produced the evidence.

Before then New Zealand had always been considered too small and isolated from other large land masses to support huge reptiles, and over the course of millions of years the land had been drowned time and again. Therefore, the thinking went, if there had been dinosaurs, their fossil traces would have been obliterated.

But Wiffen, an avid rock collector who became hooked on fossil gathering in the 1960s, did not let scientific scepticism put her off. Largely self-taught, she gradually came to the conclusion that dinosaurs may well have existed in New Zealand. Her curiosity was aroused when she discovered a 1950s geological map with these words written in the margin: 'In the Te Hoe valley the beds are partly brackish water, and contain reptilian remains . . .'. These observations had been made by Don Haw.

Situated at the end of a Hawke's Bay dirt road in the bush-clad Urewera Ranges, the Te Hoe Valley is 100 kilometres inland and at a height of 300 metres. But during the Late Cretaceous Period, when dinosaurs inhabited the area (75 million years ago), the valley was coastal, a region of plentiful food, where reptiles died and were buried in mud, and their bones preserved.

It was here in 1975 that then 53-year-old Joan and her husband Pont came across the tailbone from a 4-metre-long, half-tonne, carnivorous dinosaur – the first dinosaur to be discovered in New Zealand. The tailbone was identified as belonging to a theropod, smaller than but resembling a *Tyrannosaurus rex*.

Since then she has gone on to discover the fossils of another two types of carnivorous dinosaur, three kinds of herbivorous dinosaur, and one kind of flying reptile, similar to an *Anhanguera*, which is one of many types of pterosaur described from Brazil.

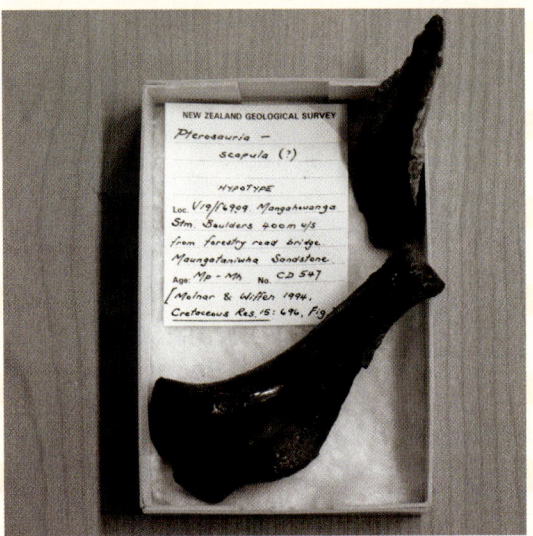

Before she unearthed these, Joan Wiffen had already earned her place in paleontological history with her discoveries of marine reptile fossils such as mosasaurs and elasmosaurs – Loch Ness-type predators, which plundered the oceans during the dinosaur period.

New Zealand's rugged terrain is bound to conceal more dinosaur fossils, but it will take someone with the dogged determination of the 'dragon lady' to find them.

i. Amateur paleontologist Joan Wiffen rocked the scientific establishment with her 1975 discovery of a dinosaur bone in inland Hawke's Bay. Before then she had unearthed this bone of a plesiosaur, a long-necked marine reptile (not a dinosaur) that swam in the oceans around New Zealand from more than 200 million years ago up until 65 million years ago, when it became extinct.
ii. A flying reptile with a 4-metre wingspan, this pterosaur ruled the skies during the age of the dinosaurs. Not a dinosaur itself, this species *Anhanguera* (pronounced 'un-arn-jeera') was a fish-eater. Although, like modern-day birds, it had hollow bones, the pterosaur was not the ancestor of birds.
iii. Dinosaur and marine reptile remains are not the only vertebrate fossils to be found in Mangahouanga Stream. Turtle bones have beeen collected and also pterosaur bones such as this one. Pterosaurs were flying reptiles that coevolved and coexisted with the birds.

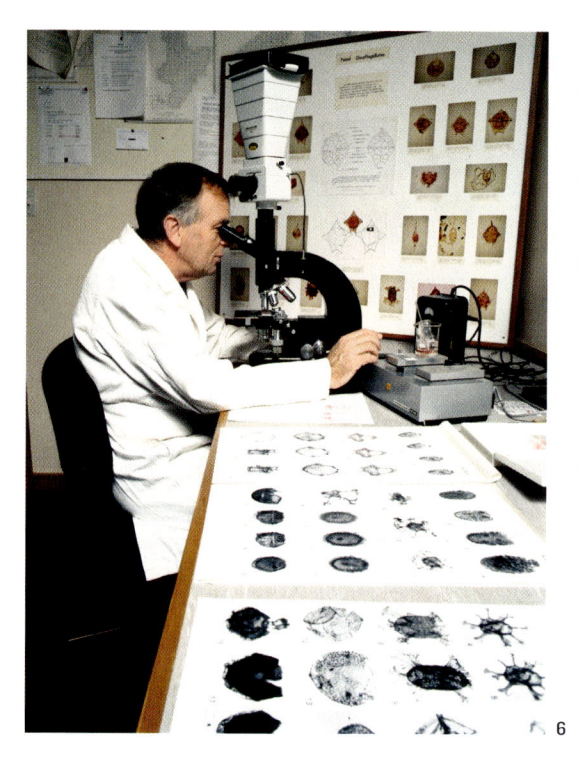

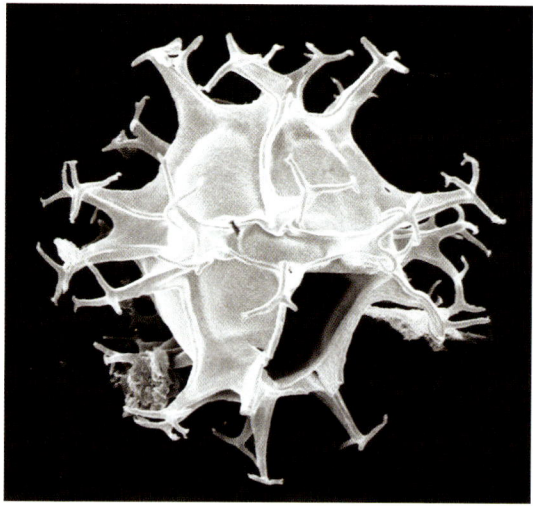

dinosaurs must have differed from those known from elsewhere in the world, such as in Gondwanaland and Laurasia. After all, they presumably would have been the result of up to 10 million years of isolation on Zealandia, and a lot of evolving can happen in 10 million years.

We can, therefore, think of the Hawke's Bay dinosaurs as Zealandian. Surely they must have been unique to Zealandia?

The Dinosaur Fossils Themselves

It is unfortunate that we know so little about these animals. However, that we know of them at all is amazing in itself, given that the sedimentary rocks in which they occur are marine, albeit marginal marine. They must have been deposited in the sea near a river mouth or beach environment in shallow water, no more than a few tens of metres deep.

The reason for expressing amazement is that dinosaurs were strictly terrestrial; that is, land-dwelling. So their remains are would not be expected to be found in sediments that were deposited on the sea floor.

Geologists refer to the rock that they occur in as the Maungataniwha Formation. The original sands and silts were almost certainly deposited in the sea near the mouth of a river.

The task of finding these fossils and extracting them from the rock is worthy of mention because it is both difficult and laborious. The fossil locality is in remote, steep, bush country within a treacherous, deeply incised, cantankerous stream. It is choked with slippery boulders and deep pools, usually located below steep waterfalls.

Joan Wiffen and her colleagues have collected specifically for vertebrate fossils. As yet, no whole skeletons have been found. Most of the fossils are at best fragmental remains of skeletons. Some are skulls (no dinosaur skulls, however) and some bones are in close association; some appear to be articulated and in correct anatomical position with respect to each other. However, most fossils are of single bones or fragments of single bones.

Many individual bone fossils have now been found and they are representative of most of the major dinosaur groups. At least six different kinds of dinosaur are recognised and they include gigantic four-legged herbivorous sauropods such as the titanosaurs, and smaller armoured ankylosaurs, bipedal herbivorous ornithopods such as the hypsilophodonts, and carnivorous bipedal theropods such as megalosaurs and allosaurs.

Not one of the dinosaur fossils found so far from New Zealand warrants identification to species level. In other words, the bones cannot be related to any particular known species of dinosaur. The fossil evidence available is too meagre and the individual bones or bone fragments permit only a generalised identification. For instance, the very first dinosaur fossil found is a small vertebra. It is distinctive enough, on the basis of its shape and size, to be certain that it must come from near the end of the tail of a theropod

6. Graeme Wilson is a palynologist who has pioneered the study of dinoflagellates. These are organic-walled reproductive products (resting cysts) of a major group of marine plants (algae). In many ways they are comparable to pollen and spores of terrestrial plants, yet they are very different in shape and structure. Dinoflagellate research has greatly enhanced our understanding of the Late Cretaceous-Eocene marine record of Zealandia.

7. A typical fossil dinoflagellate cyst photographed using a scanning electron microscope. The dimensions of this fossil are about 60 microns in height and 30 microns in width. This is a specimen of *Spiniferites ramosus* of Late Cretaceous age, about 88 million years old.

8. The Chatham Islands as seen from space. The longest east–west dimension in northern Chatham Island is 65 km, whereas the longest north–south dimension is 50 km. Home to about 700 people, the Chatham Islands are the eastern-most part of New Zealand, located about 850 km due east of Christchurch.

9. Only three areas in New Zealand have so far thrown up dinosaur fossils: Hawke's Bay, Port Waikato and the Chatham Islands. On this wave-cut platform on Chatham Island within a formation called the Takatika Grit, dinosaur fossils include a tiny claw and finger, spinal, foot, and leg bones from an unknown variety of two-legged meat-eaters known as theropods.

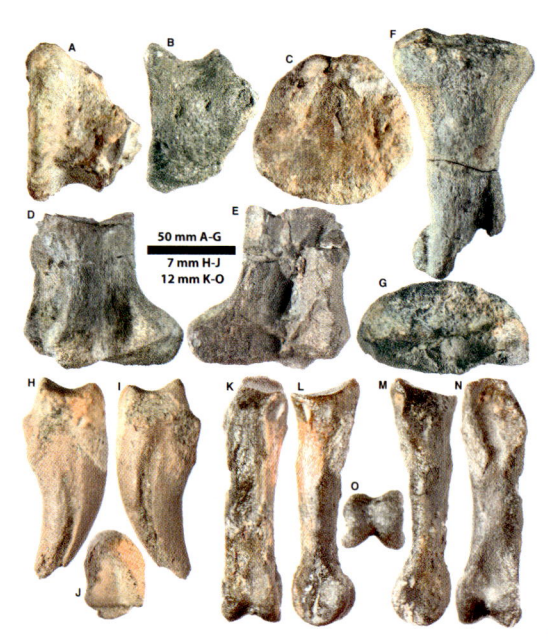

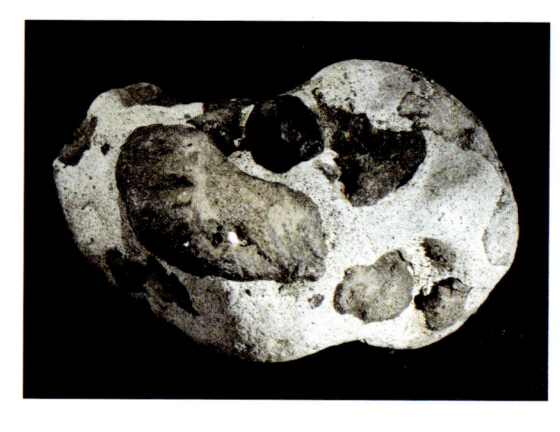

dinosaur, and it is closest in shape to those of allosaurs. This is all that can be said without more fossil material.

Similarly, at least two fragmentary bones have been found that are attributed to ankylosaurs. One is a segment of thin, curving rib-bone, with a ledge developed along its length. It is very distinctive and is characteristically ankylosaur. The ledge helped support the armour plate! But trying to relate this bone to any known species of ankylosaur is almost impossible without more fossil evidence.

However, in light of our geological knowledge of the history of Zealandia, we can be reasonably certain that these fossils must indeed be representative of species of dinosaur that were unique to Zealandia. In this regard, there is no reason why they should not be formally described and attributed unique species names of their own, no matter how scrappy the fossils are.

Apart from the Hawke's Bay locality, only two other dinosaur localities (in the Chatham Islands and near Port Waikato) are currently known from New Zealand, but no doubt others will be found. These other two can be attributed to Gondwanaland and may be regarded as Gondwanan. Only the Hawke's Bay locality is Zealandian.

Gondwanan Dinosaurs on the Chatham Islands

In 2006 five dinosaur bones were described by a research group of Melbourne-based vertebrate paleontologists, led by Jeffrey Stilwell and Chris Consoli, from a rock formation exposed in a shore platform on the northern coast of Chatham Island. Apart from one vertebra, they are all foot or hand bones, but they can all be related to theropod dinosaurs. At least one of them belongs to a large animal about 6 metres long, similar to an allosaur.

They are interpreted to be of Late Cretaceous age, 100–90 million years old, on the basis of associated fossils. They predate the separation of Zealandia from Gondwanaland and must therefore be regarded as Gondwanan dinosaurs.

However, they occur within an unusual and very distinctive sedimentary rock that is of much younger, Early Paleocene age, about 63–62 million years old. Once again, we can be precise here because of associated dinoflagellate fossils. Such is the power of microfossils, and especially fossil plankton.

This means that the Chatham Island dinosaur fossils are what geologists refer to as 'reworked'. They are preserved in Cretaceous sediments that significantly predate the Cretaceous–Paleogene mass extinction event (65 million years ago), and yet they are in sediments that postdate this cataclysmic event. They have subsequently been eroded out of the original sedimentary formation that they were deposited in and have become incorporated as pebble- or cobble-sized components within much younger sediments, in this case the Takatika Grit.

We do not know exactly where the original source rock for the dinosaur fossils was or is on Chatham Island, but it was probably nearby. The fossil bones may be a sort of 'lag deposit' that remained after erosion of the 'source sediment'. In other words, they represent the remnants of a pre-existing formation, now long-since destroyed.

Suitable rocks, however, do occur on Pitt

10. Dinosaur bones from the Chatham Islands. These are some of the fossil bones collected from the Takatika Grit of Early Paleocene age, 63–62 million years old. The bones must be older than the sedimentary rock that they occur in because the dinosaurs became extinct 65 million years ago! The bones are probably derived from older underlying sediments, such as the Tupuangi Formation. They are best regarded as Gondwanan, rather than Zealandian. These fossils were discovered by Jeff Stilwell, Chris Consoli and colleagues.

11. A cobble derived from the dinosaur-bearing Takatika Grit on Chatham Island with conspicuous phosphorite nodules. The dark green grains in the matrix are glauconite, an authigenic, potassium-rich mineral that grows on the sea floor. This unusual sedimentary rock is of Early Paleocene age (63–62 million years old) yet identical sediment can be found on the sea floor of much of the Chatham Rise today. In this respect the geology of the Chatham Islands provides an on-land window on the geology of the Chatham Rise. The cobble is 100 mm x 65 mm.

12. Tupuangi Formation, Waihere Bay, Pitt Island. These are the oldest sediments in the Chatham Islands and are river sediments of Cretaceous age, between 100 and 90 million years old. They accumulated on the eastern margin of Gondwanaland before Zealandia rifted away. They are remarkable because they have remained almost undisturbed, bar a few volcanic eruptions nearby (Mangere Volcano) 6 million years ago. Fossil forest floors complete with tree stumps and roots are preserved within these rocks and potentially they may contain Gondwanan dinosaur fossils.

Island, almost 60 kilometres to the south-east. This is the Tupuangi Formation that is best exposed in Waihere Bay on the west side of Pitt Island. This formation would be a good place to search for dinosaur fossils, although no bone fossils have yet been found there. The formation includes fossil soils complete with fossil tree stumps and gravels (conglomerate) deposited by rivers.

The Takatika Grit formation is observed only on northern Chatham Islands and is flat-lying and not very thick at less than 6 metres. It rests directly on schist basement and is rich in authigenic minerals. It is called the Takatika Grit and is named, complete with spelling mistake, after the nearby Tahatika Stream. It is truly gritty, with rock fragments and quartz derived from the underlying schist.

Authigenic means 'formed in place'. In other words, these are minerals that grow *in situ* within sediment on the sea floor from chemical nutrients dissolved in sea water. The perfect setting for the growth of such minerals is on submarine rises such as the Chatham Rise, where deep, cool, nutrient-rich waters are forced up into shallow, warmer water.

Potassium is one of the elements in high concentration. Given the right conditions, it precipitates from sea water to form the mineral glauconite. Glauconite can grow in great abundance to produce a sand, and by virtue of its strong green colour, it forms 'green sand'. Greensands are surprisingly common in New Zealand sedimentary rocks of Paleogene to early Neogene age (65–15 million years ago).

Phosphate also precipitates to produce smooth, irregular-shaped, chestnut-brown rock masses referred to as 'phosphorite nodules'. They can be as big as very large potatoes, and look like them too.

From oceanographic accounts, the sea floor of the Chatham Rise today looks remarkably similar to what can be seen on land at Tioriori on Chatham Island. Yet the Takatika Grit is 63–62 million years old!

The Chatham Rise has been explored, but not exploited, for its rich reserves of potassium-rich glauconite and more particularly for its phosphorite nodule resource. Not only is the phosphorite rich in phosphate, but it also contains unusually high concentrations of uranium and gold. The phosphate is of special interest because trials have shown that the phosphate is in a form that is readily amenable to uptake by plants. So if it could be easily mined, all that would then be necessary would be to crush it and spread it directly on pasture as a fertiliser. It is allegedly of better quality than superphosphate.

Furthermore, this sea floor on which the Takatika Grit accumulated was probably remote from land, as is suggested by the abundance of authigenic minerals. They indicate an almost total absence of sediment supply from a terrestrial source. Most of the Takatika Grit is authigenic and the rest is biogenic (of biological origin). Associated marine fossils indicate significant water depths of at least 200 metres and probably 400 metres.

At Tioriori, in among the abundant phosphorite nodules can be found relatively common fragments of fossil bone scattered among more common fossil sponges that have shapes similar to bones. The distinction can be very confusing to the uninitiated. The Takatika Grit is a most unusual setting in which to find dinosaur bones, on a relatively deep marine sea floor. In this regard, the dinosaur fossils can only be explained as having been reworked from some other source.

Other Cretaceous fossils have been found in the same formation, including mosasaur bones (marine reptiles) and ammonites, nautiloids, gastropods and bivalves (clams). It is presumed that all of these old fossils have been reworked into younger sediment by erosion of older sediments.

No matter how these discoveries and observations are interpreted, they are of great interest and fascination. This is only the third dinosaur locality so far discovered from New Zealand, and the second Gondwanan dinosaur locality. Although they must be Gondwanan, they occur in Zealandian sedimentary rocks. How confusing!

Potential for Zealandian Dinosaurs

In theory, there is considerable potential for finding Zealandian dinosaur fossils in New Zealand. Suitable continental sedimentary rocks that accumulated on land, not in the sea, are relatively widespread, especially in the South Island. These are pebble conglomerates, sandstones, siltstones, mudstones and coal measures of Late Cretaceous age. These rock sequences are the focus of attention for the coal-mining industry in the South Island, especially on the West Coast, Southland (Ohai) and Otago (Kaitangata).

The original sediments would have been gravels, sands, silts, mud and peat that formed in low-relief river systems, meandering across a subdued landscape with attendant swamps and lakes. The landscape would have been similar to Australia today, only much wetter and lusher.

It is perhaps a little surprising that other dinosaur localities have not yet come to light, but only a few people are out there looking.

A Gondwanan Jurassic Dinosaur from Port Waikato

A tantalising, single fragment of a small bone, identifiable as part of a finger or toe of a small theropod dinosaur, was found by Brendan Hayes in rocks of Late Jurassic age from near Port Waikato. These rocks are about 145 million years old, so they are 70 million years older than the dinosaurs of Hawke's Bay. Furthermore, the Jurassic dinosaur that owned this bone would have been

Gondwanan not Zealandian. This is the only dinosaur fossil currently known from mainland New Zealand that is attributable to Gondwanaland.

It is possible to make a distinction between Gondwanan dinosaurs and Zealandian dinosaurs. Any older than 83 million years are Gondwanan, and any younger are Zealandian.

There is some potential for finding Gondwanan dinosaurs. Terrestrial and fluvial (river-deposited) sedimentary rocks that are older than 83 million years old are present in some places in New Zealand. They range in age from Middle Jurassic to middle Cretaceous, so from about 175–100 million years.

As mentioned above, one of the more promising localities is on Pitt Island in the Chatham Islands, where the Tupuangi Formation forms basement to all younger rocks. It is widespread on Pitt Island, particularly the northern half of the island, but it is best exposed in Waihere Bay on the west coast.

Other potential localities include the Catlins area of Southland (Middle Jurassic) and the Port Waikato area south of Auckland (Late Jurassic).

The Demise of the Dinosaurs

It is widely accepted within the scientific community that whatever event is responsible for the mass extinction at the end of Cretaceous time, it was also responsible for the extinction of the dinosaurs. The consensus is that the Earth was struck by a large meteorite, and widespread extinction occurred as a consequence of a series of disastrous, knock-on environmental effects.

It is an amazing story and is extremely well supported by geological evidence. First advanced in 1982 by Walter Alvarez, this explanation was finally accepted in 1990, with the discovery of the crater created by this meteorite on impact. This is the Chicxulub Crater, located in Mexico, partly in the Gulf of Mexico and partly on the north-eastern edge of the Yucatan Peninsula. The crater is almost 200 kilometres in diameter and is completely buried by younger sediments that are less than 65 million years old. No wonder it was hard to find!

It has been determined that a crater this size must have been generated by a meteorite about 10 kilometres in diameter: a substantial body, yet fairly insignificant compared to the diameter of the Earth (12,720 kilometres). Nevertheless, the meteorite proved immensely destructive. It was travelling so fast that it would have taken less than two seconds to punch through the Earth's atmosphere.

Unluckily, the target included limestone ($CaCO_3$) and thick evaporite deposits, rich in salt (NaCl) and gypsum ($CaSO_4$). The pressure wave generated by frictional effects as the meteorite passed through the atmosphere must have been so intense that temperatures would have exceeded those at the surface of the Sun. On impact the meteorite was instantaneously vapourised and along with it a huge volume of the substrate that it smacked into.

It sounds like fantasy, but all this is known from the chemistry of a distinctive layer of dust that was undoubtedly generated by this event. This layer, known as the 'boundary clay' because it relates to the Cretaceous–Paleogene boundary (commonly referred to as the Cretaceous–Tertiary boundary), formed as a blanket that slowly mantled the Earth's surface as a result of a vast envelope of fine, particulate dust and aerosols that engulfed the globe and blacked it out. The boundary clay is well preserved in many places around the globe, especially in sedimentary basins, both on land and in the oceans. It is best preserved in quiet settings such as deep marine sediments.

From a detailed yet simple calculation and modelling of the fallout products from the impact, it has proved relatively easy to work out the original size, composition and speed of the meteorite. It has also been possible to figure out what killed the dinosaurs. After all, they were not all amassed on the one spot in the Yucatan Peninsula at a major gathering when the meteorite struck. They died out primarily as a result of darkness: a total black-out caused by a dense, thick, pervasive fog of atmospheric dust.

The impact threw so much fine particulate material and chemical aerosols into the atmosphere that sunlight was prevented from reaching the surface of the

13. Meteorite arriving on Earth. An artist's impresion of the Chicxulub Meteorite impacting with the Earth. About 10 km wide, it landed on the north side of Yucatan Peninsula, Mexico, vapourising on impact and generating a crater some 200 km wide. Its effects on the Earth's atmosphere were so profound that no light could reach the surface. Photosynthesis was stopped for a period of several months resulting in collapse of the food chain and extinction of most large animals, including the dinosaurs.

Earth. This situation is thought to have lasted for at least three months. With no light, the base of the food chain failed. All photosynthesis would have ceased and therefore all plants would have died or shut down, their leaves useless. The rest of the food chain would have followed suit. Single-celled algae (plants) form the base of the food chain in the oceans. They perished as well, with the resulting collapse of all higher organisms.

Plants did survive by virtue of storage organs and seeds. Animals survived too, but only those that could burrow and hibernate, or survive as scavengers. Many would have perished from toxins, not to mention a host of other post-apocalypse ravages associated with prolonged darkness, such as the inability to see and find food, and the intense cold of an endless night.

It is estimated that all animals with a body weight in excess of 25 kilograms perished to extinction. This explains the demise of large animals both on land and in the sea. Smaller animals survived, including many mammals, and no doubt some small dinosaurs. After all, in strict biological terms birds are regarded as extant dinosaurs, because they conform to the scientific definition of a dinosaur: an animal with a diapsid skull (having two apertures in the skull behind the orbit of the eye) and legs beneath their bodies. This definition explains why the flying reptiles such as pterosaurs, and the marine reptiles such as the mosasaurs, are not dinosaurs: they have legs at the sides of their bodies like lizards, and different skull structure.

The Cretaceous–Paleogene Boundary in New Zealand

Zealandia was not immune to the catastrophic visitor to Mexico 65 million years ago. The famous 'boundary clay' has been discovered in New Zealand. Furthermore, it has been recognised in well-known Zealandian sedimentary sequences in the South Island, in both terrestrial and marine rocks. It occurs within coal measures on the West Coast near Greymouth, and in limestone in Marlborough between Blenheim and Kaikoura.

It is characterised by tell-tale chemistry and not least by the presence of elevated levels of iridium. This strange platinoid metal just happened to be a significant chemical component of the killer meteorite. What a stunning calling card, and so easy for forensic geologists to trace!

To See Zealandia

What did Zealandia look like prior to sinking? What kind of landscape existed? These are interesting questions. For most of us, it is hard to visualise what it looked like, because modern-day New Zealand is so very different from Zealandia. The reason why is this: New Zealand as we know it is the product of vigorous plate collision over the past 20 million years. Our landscape has been highly modified since Zealandia.

Almost everywhere we look there is evidence of this collision. The flat plains are all very young, the product of floods of sediment derived from rapid erosion of rising mountainous hinterland. Or in the case of the central North Island, much of the flat land is in fact due to smoothing of the topography by thick volumes of volcanic ash. Most river and stream valleys in New Zealand are surprisingly freshly cut and actively cutting down.

The most obvious features are mountain ranges. Our rugged landscape is all about relentless deformation. The mountains have been and are being pushed up. And as they are pushed up, the rivers cut down and add extra relief to already high and steep topography.

Conclusion

Having addressed what is known of our six Zealandian dinosaurs and their demise 65 million years ago, we need to consider Zealandia's subsequent history. It continued to slowly sink for another 40 million years until about 25 million years ago. Then everything changed and Zealandia was transformed into New Zealand, but in order to make sense of this process, an appreciation of faults and earthquakes is necessary. Hence the next chapter.

10/ Crust Busters: The Tyranny of Faults

The colossal forces associated with plate collision are directional and strong. They are not unlike the crushing forces of a carpenter's vice or a nutcracker. These tectonic forces cause the Earth's crust to break, and the fractures or lines of breakage are called 'faults'. They can be thought of as lines of weakness within the rock, or as lines of failure.

When a fault is mapped, it is always shown as a line. Faults may appear to be haphazard, but they are never random. There is always a hidden control or reason for their presence and their distribution pattern. Because tectonic forces are directional, the resultant faults form in a systematic and predictable manner. Accordingly, we have learnt to understand them. We can work out why they are where they are, and if we are lucky, when the breakage occurred.

Kapiti Island

Upper Hutt
Lower Hutt
Wellington City

1. A number of active faults lie within this field of view, including Shepherds Gully, Ohariu and Wellington. They are all transpressional; that is, when they break or rupture, they move horizontally and also vertically (the ratio is about 2:1). They all have different histories and rates of movement, yet they are all the result of compression due to collision between the Pacific and Australian plates. They all cut the full thickness of the crust some 25 km. The Wellington Fault is enabling the western hills to climb up over the harbour! It is part of a semi-continuous feature that can be traced from south of the Wellington coast to the Bay of Plenty.

Let us explore faults further. In order to understand what Zealandia looked like, it is first necessary to understand our present landscape. In order to 'see' what the Zealandian landscape looked like, we need to subtract the effects of the past 23 million years of carnage on our faults. Faults are important to any investigation of the geological history of New Zealand. They define the edges of all the components that make up the visible New Zealand landscape.

If you think of New Zealand as a great fish, then the scales are crustal blocks and the edges of the scales are faults. However, this analogy only works so far, as some faults are much bigger than others. Generally, when a geologist finds a fault, the first question is: where is 'The Big One', the master structure that has caused this smaller one to form? Most faults are 'noise' to much bigger faults. Even the really big faults that dominate the eastern North Island, such as the Wellington Fault and the Wairarapa Fault, may be thought of as secondary features to a much bigger structure, the sheet-like surface of dislocation between the down-going or subducting Pacific Plate, and the over-riding Australian Plate.

Fault Lines

Why are faults shown as lines on geological maps? Think of faults as planar surfaces along which movement has occurred. Usually, fault surfaces are almost vertical or inclined at a steep angle. When one surface intersects another surface, there is a unique line of intersection. Similarly, when a fault intersects the surface of the land, which is almost horizontal, the junction where they meet defines a line, and lines are easily depicted on maps.

Faults are planar surfaces that develop within the rocks of the Earth's crust. They enable discrete volumes of crustal rock to move past one another. Colossal tectonic forces operating within the Earth require changes in volume or shape of the Earth's crust, and to facilitate this process, movement is required and it takes place on faults. They accommodate the stresses of tectonic forces and permit relief of accumulated strain. They are a normal response to fluid processes that are moving unseen and at depth beneath a rigid skin. The skin is thin and easily deformed. So, too, is the Earth's crust.

There are faults everywhere in New Zealand, and there is a great variety of them. Some are small, some are large, and some are active, which means they have moved within the past 100,000 years.

Normal Faults

Faults are characterised in terms of how they formed. For instance, if the crust is being stretched and pulled open, so-called 'normal faults' form, whereby one side ramps down the other side so that the net result is an extension of the Earth's crust. It has literally 'grown' extra surface area and has expanded. To do this effectively, the major faults cut right through the full thickness of the crust.

Normal faults rarely exceed 60–70 kilometres in length, and the earthquakes associated with their movement or rupture rarely exceed 7.5 in magnitude. Normal faults are generally straight and break cleanly with relatively small foreshocks and aftershocks being generated. The reason for this has to do with rheology of the crust; that is, its strength, and the way it moves and 'flows'. The thickness of the brittle crust and the curvature of the Earth are

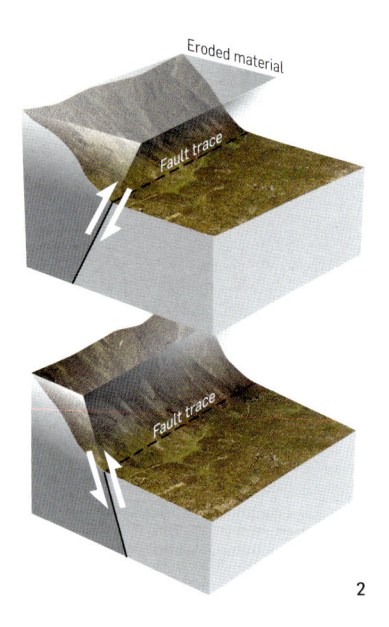

2. Most faults are described as 'normal' or 'reverse'. In a normal fault, when the Earth's crust is stretched or extended, the block on one side of the fracture slips down below the other, as happened on the Edgecumbe Fault during the Edgecumbe Earthquake. In a reverse fault, the crust is squeezed under compression and the block on one side of the fault is pushed up over the top of the other. The net result is a shortening or reduction of the land surface. The 7.8 magnitude Murchison Earthquake in 1929 was caused by a reverse fault, as was the mighty 1855 Wairarapa Earthquake.

3. At 27 km in length, the Paeroa Fault is the largest fault of the modern Taupo Fault Belt. One of New Zealand's most active geothermal fields, Te Kopia, extends 2.5 km along the 220-metre-high scarp of the fault. This is a normal fault, a mechanism that permits the crust to break under tension due to to stretching or rifting, with the net result of extension. In other words, normal faults allow the surface of the Earth to grow.

4. A 'normal' fault that had not been identified before it made its dramatic appearance in 1987 was the Edgecumbe Fault, which created a rupture 7 km long that in some places was 4 m deep and 2 m wide. This view of McCracken Road is instructive because it shows an unaffected power pole supporting a heavy transformer. What it means is that the displacement on this segment of the fault was clean, with almost no attendant ground shaking.

constraining factors. In other words, given a particular crustal thickness, there is a natural limit to how long a normal fault can break the Earth's surface.

In New Zealand, the best place to see normal faults is within the Taupo Volcanic Zone. This is our very own active rift zone where the crust is being stretched in an east–west direction. Normal faults can be found on either side of the Taupo Volcanic Zone. Some of these faults have produced spectacular fault scarps, such as the Paeroa Fault south of Rotorua.

Edgecumbe Fault

Not so spectacular to look at, but entrenched in the minds of many New Zealanders, is the Edgecumbe Fault. Prior to it moving and making its presence felt in 1987, it had not been recognised. Its surface expression is, or at least was, cryptic. Despite this understandable oversight, it is the perfect example of a normal fault. It last ruptured and moved in the Edgecumbe Earthquake, which was 6.3 in magnitude and happened at 1.42 p.m. on Monday, 2 March 1987. The down-dropped side of the fault, to the west, dropped by up to 2 metres, and the length of the fault rupture was about 7 kilometres. A number of people were injured and there was lots of damage to property, but, as luck would have it, there were no deaths.

This was the most recent devastating earthquake that New Zealand has experienced. We are overdue for another. From all available historical and geological evidence, there is a strong suggestion that New Zealand has been let off lightly in the

past 100 years. Such is nature: it is not so regular. On average about six earthquakes greater than magnitude 6 are recorded from the New Zealand region each year.

Reverse Faults

'Reverse' faults are the opposite of normal faults, forming as a result of compression or squeezing, with the net result of a reduction in surface area. One side of the fault is literally ramped up over the other side. These faults can be very long and sinuous. When they rupture, they tend to be accompanied by lots of earthquakes, both foreshocks and aftershocks, reflecting adjustment along an irregular, high-friction surface.

Many of New Zealand's largest and most active faults are reverse faults, including the Alpine Fault and the Wairarapa and Wellington faults and their

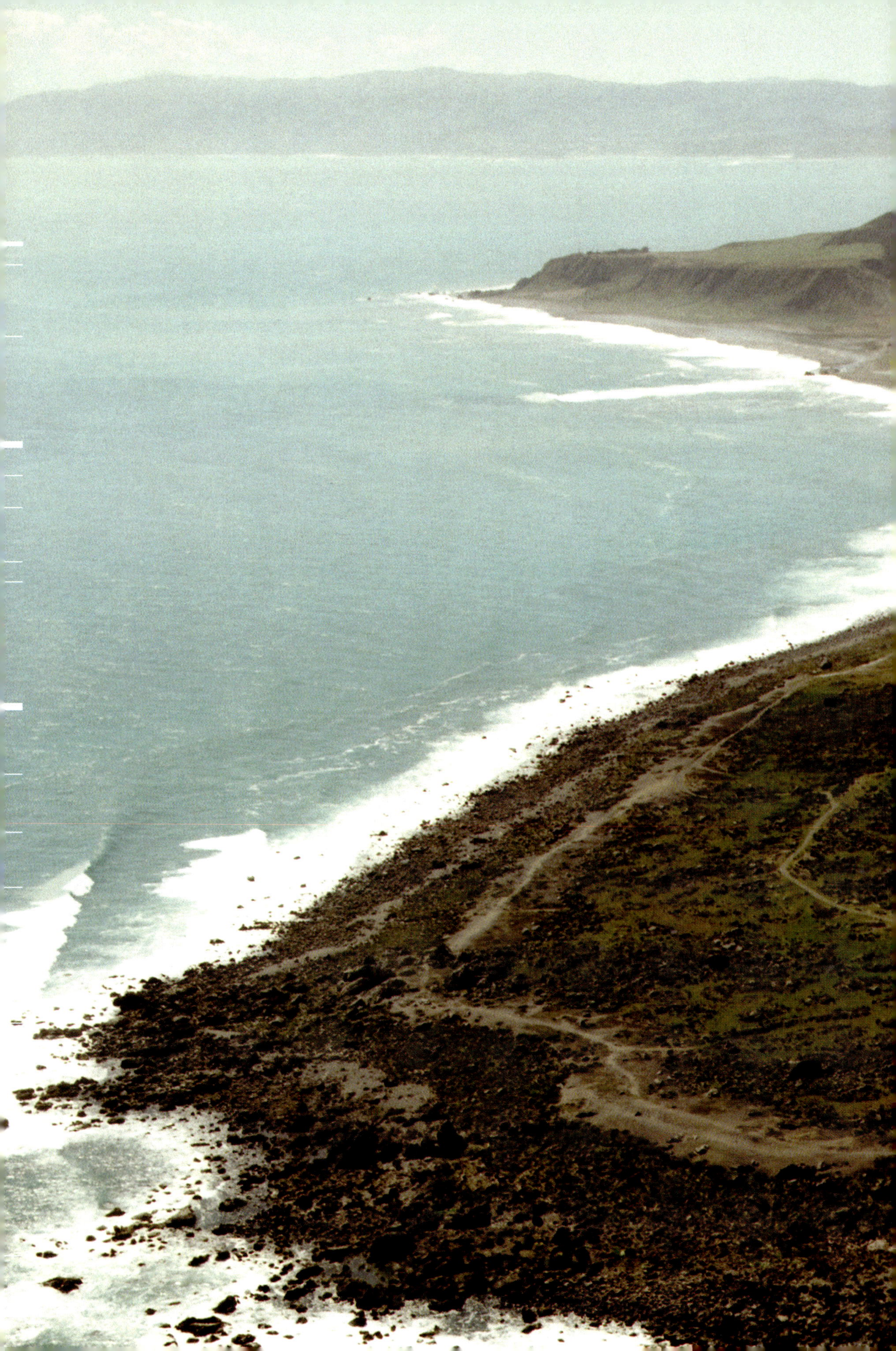

5. Spectacular proof of New Zealand's restless earth are these five stranded beach ridges along the coastline at Cape Turakirae, east of Wellington, providing a continuous record of upheaval over the past 7000 years. They show that the area has experienced successive earthquakes in excess of magnitude 8. In 1855 movement on the Wairarapa Fault during the Wairarapa Earthquake raised the beach 6.5 metres. It seems that every time this fault moves, about every 1000 years, it jumps vertically 6 m or so. But much more dramatic is its sideways movement. In 1855 it stepped 17 m!

northerly extensions. But they are not entirely reverse faults; they have an extra characteristic. These faults are moving sideways or laterally, almost twice as much as they are vertically. They are referred to as 'transpressional' faults, for the simple reason that they are responding to tectonic collision forces that are operating across the fault at an acute angle rather than at right angles to it.

The Wairarapa Fault

When the Wairarapa Fault last moved, it moved up to 17 metres laterally and up to 6.5 metres vertically, and it ruptured over a distance of at least 156 kilometres. It did so when the Wairarapa Earthquake struck at 9.17 p.m. on Tuesday, 23 January 1855. Most damage was caused that balmy summer's evening within the brief minute between 9.17 p.m. and 9.18 p.m. This was the last great earthquake to strike New Zealand, and it was widely felt over much of central New Zealand, with thousands of aftershocks over a period of many months.

Surprisingly, we now know of more than 20 rupture events on the Wairarapa Fault. This number of events provides a reasonably robust statistical basis for forecasting future events, unlike the Wellington Fault for which we know of only two events. The recurrence interval or repeat period is about every one thousand years. So, in the case of the Wairarapa Fault, for once we can rest easy with a reasonable degree of certainty that nothing is likely to happen for almost a thousand years.

In terms of magnitude, it is estimated to have been at least 8.2, based on all known aspects of the earthquake and its effects. There were no seismographs in those days and therefore no instrumental records. Amazingly, only one death was reported in Wellington, that of Baron von Alsdorf, hotelier and owner of the finest and newest hotel on Lambton Quay. Allegedly, he bled to death after being wounded by flying glass from a large, smashed mirror. At the time, about 3200 people lived in Wellington, another 1800 in the Hutt Valley and about 1500 in the Wairarapa.

The Wairarapa Fault is a transpressional fault that is permitting the crust on the eastern side, including the Wairarapa Plains, to slide south with respect to the axial Rimutaka, Tararua and Ruahine Ranges. At the same time, it is allowing the axial ranges on the western side of the fault, to climb up over the Wairarapa Plains.

If you squeeze a sheet of corrugated cardboard, the ridges go up and the valleys go down. In a sense, this is exactly what is happening across the southern part of the North Island. The major valleys and/or depressions are being driven down and the major ridges and ranges are being driven up. The east–west dimension is trying to become shorter as plates collide. This analogy is complicated by the considerable component of 'drag' or 'shear' imposed by oblique transpression. This is why the active faults in the eastern part of the North Island, the Wellington area included, are all moving sideways at least twice as fast as they are moving vertically.

The Wellington Fault

When the Wellington Fault last moved, it did so 3–4 metres laterally and 1.5–2 metres vertically. It last moved in the so-called Hauwhenua Earthquake of about AD 1450. Only two rupture events or movements of the Wellington Fault are known at present, but when it moves, it is estimated that the earthquake is in excess of magnitude 7.5, and it is thought to move every 400–700 years. Some geologists have claimed that there is a 10 percent chance of the Wellington Fault moving within the next 50 years, but they have been making this pronouncement for almost as long! The reality is that we have no idea when it will next move, so it is therefore important to be prepared at all times.

The Wellington Fault is especially familiar to most Wellingtonians because it is so conspicuous. The Hutt–Kaiwhara Road snakes along a very pronounced erosional scarp developed along some 50 kilometres of its length. The western side of the fault is trying to climb up over the eastern side. In other words, the 'western hills' of Wellington are trying to rise up over Port Nicholson (Wellington Harbour) and the Hutt Valley. This effect creates a 'free edge' that collapses easily and is quickly eroded away. The result is a pronounced, steep-faced bevel that many people misinterpret as the fault. The rock that makes up the 'free edge' of the uplifted fault block to the west side of the fault is greywacke, but it is in very poor condition. It is intensely crushed and fractured and has the consistency of Weetbix. No wonder it fails.

To make matters worse, the Wellington Fault bends as it cuts its way from Wellington to Upper Hutt and beyond. When it moves, it has to accommodate movement around this bend, which is very difficult to do without intense shaking and attendant land-sliding effects. This explains why the erosion scarp of the Wellington Fault is such a profound feature, set well back from the actual location of the fault, which is tens of metres offshore in Wellington Harbour

The Rootless Eastern North Island

Much of the eastern North Island may be thought of as the feather-edge of the Australian Plate. This continental crust forms an increasingly more deformed wedge that thins to the east. No wonder transpressional plate collision is pushing it around: it simply isn't strong enough to stand up for itself. In fact, the entire eastern strip of the North Island may be thought of as rootless. Everything east of the main axial ranges along a line drawn from Whakatane to Wellington, is entirely detached from the Australian Plate. This is the triangular wedge of crust on the Australian Plate that is cradled between the Wellington–Whakatane Fault zone to the west, and the subducting Pacific Plate beneath.

Our Worst Earthquake

The largest earthquake to strike New Zealand since 1855 was much worse in terms of the number of victims, yet much less powerful. This was the 1931 Hawke's Bay Earthquake. It struck at 10.47 a.m. on 3 February and was 7.8 in magnitude. It lasted for 30–35 seconds and in that time triggered a series of events that were to bring the Hawke's Bay region to its knees. By far the worst effect was fire.

At least 257 people died as a result of this earthquake, compared with a total of only five or six in the 1855 earthquake. In Napier, 161 people lost their lives, 93 in Hastings and three in Wairoa. Some 400 were seriously injured, and at least 2000 injured in total. Some 11,000 people had to be evacuated. Within the first 12 hours of the earthquake there were 150 aftershocks, and within the following 14 days some 500 aftershocks, including a magnitude 5.7 earthquake on 4 February, a 6.3 earthquake on 9 February, and a 7.3 earthquake on 13 February. So what happened?

The Poukawa Fault

Modern analysis has shown that the earthquake ruptured the Poukawa Fault. It was initiated more or less at the northern end of the fault (as mapped at the surface) beneath Hastings and propagated in a south-west to north-east direction towards Napier and on up towards Wairoa. There was a major dislocation on the fault plane at a depth 5–20 kilometres below the surface. The actual displacement is estimated to have been a maximum of 8 metres sideways and 8 metres vertically. To put this in perspective, the land to the east of a line drawn between Hastings and Wairoa moved to the south-west, and the land to the west of the same line tried to climb up over the land to the east of it. A classic double whammy effect!

This can be achieved only on a reverse fault that is moving both sideways and vertically. So the Poukawa Fault is a steeply inclined surface that dips or deepens to the west beneath the coastal plains of Hawke's Bay. Until the 1931 earthquake, geologists had no knowledge of the northward existence of the Poukawa Fault beyond Hastings.

For all that the fault dislocation at depth was considerable, the actual deformation of the land surface was much less dramatic. Maximum uplift was 2.5 metres and maximum subsidence was only 1 metre. Nevertheless, the surface area affected was 90 kilometres in length and 17 kilometres in width.

On the basis of current geological understanding, the Poukawa Fault has ruptured in a similar way to the Hawke's Bay Earthquake every 3000–5000 years. However, there is no room for complacency: there are at least 20 active faults within 100 kilometres of Napier. All of these faults have the potential to move catastrophically. They are all within the leading eastern edge of the Australian Plate and the rocks involved are of continental crust. They are within the plate boundary collision zone and responding as best they can to tremendous transpressional forces. Whereas the Australian Plate is trying to move northwards, the Pacific Plate is trying to move westwards. The crust is being squeezed in such a way that the land surface is being forced both upwards and sideways. Accordingly, the crust is being bust: it is breaking up.

If we explore this idea further, it becomes apparent that Hawke's Bay is just part of a much larger picture. The entire eastern part of the North Island is in fact rootless. The subducting Pacific Plate is a mobile surface behaving like an extremely wide but slow-moving descending escalator. The eastern North Island is like a large load that is being carried along and rather unceremoniously off-scraped against more resistant rocks to the west along large bounding faults such as the Wellington, Wairarapa, Mohaka and Whakatane faults. The eastern North Island is like a paua being shucked of its shell.

Blind Faults

The Poukawa Fault can be regarded as a 'blind' fault, a fault capable of moving at depth with devastating effects at the surface but surprisingly little actual visible and/or mappable offset displacement. A modern example is the fault responsible for the Bam Earthquake that struck southern Iran on Boxing Day 2003. It was only magnitude 6.5 yet was responsible for 50,000 deaths. This was yet another transpressional fault, but in this case it was associated with collision between continental crust on the African Plate and continental crust on the Asian Plate.

Blind faults are of concern. Their recognition is difficult because they are not obvious at the surface and are, therefore, undetected and unmapped. Modern assessment of the 'earthquake hazard' for New Zealand is largely based on what we know of active faults with demonstrable surface expression. Therefore, hazard assessment is almost certainly underestimated. There are likely to be blind

6. At 650 km long, from Blenheim to Fiordland, the Alpine Fault is by far the longest of New Zealand's numerous faults and can clearly be seen from the air and even more so from space, but Earth-bound geologists did not recognise it until 1941. Harold Wellman and Dick Willett were the first to recognise it. They demonstrated that the distribution of two major rock types, schist to the east and granite to the west, demarcated a major geological boundary so sharp that it could only be a fault. Furthermore, they were able to show that it had been dislocated by at least 460 km. Here it is coursing its way northwards along the Cascade and Jackson valleys in northern Fiordland.

7. The ruler-straight line of the Alpine Fault running up the West Coast of the South Island marks out the Australian from the Pacific Plate. Picked out by the snow line, the fault shows up as an altitudinal feature, with the uplifted eastern side (Pacific Plate) snow covered, and the low lying western side (Australian Plate) snow free.

faults beneath much of the more extensive flat lands of New Zealand, such as the Canterbury Plains and other areas dominated by expanses of alluvial gravels.

Lessons from Kobe and Bam

To put the Wellington Fault hazard in perspective, the magnitude 6.9 Kobe Earthquake that struck the Kobe area of Japan on 17 January 1995 was associated with the rupture of the Tajima Fault on Awaji Island, and it was transpressional, moving just 75 centimetres laterally and 75 centimetres vertically, yet it was responsible for almost 5500 deaths.

GNS Science and GNS GeoNet collect and analyse data relating to all earthquake activity in New Zealand and develop earthquake hazard assessments for all parts of New Zealand. They predict the level of risk based on estimates of severity of seismic shaking and how often earthquakes will occur. They take into account all known active faults in any given area, and the nature of the ground. We now realise that these assessments must all be underestimating the true picture, the real hazard, because blind faults are not included. How can they be? And yet they clearly exist!

Size and Distance Matter

The Bam Earthquake and the Kobe Earthquake killed similar numbers of people, yet they were very different in magnitude.

There are two ways of measuring the strength of an earthquake: magnitude and intensity. 'Magnitude' is a measure of how powerful or energetic an earthquake is. It is rather like the power of a light bulb. How intense the light is depends on how close you are to the source, the bulb. The same is true of earthquakes: the closer you are to the source, the more intense it will feel and the more damaging it will be. The Richter Scale is a measure of the power released in the earthquake (like the wattage of the bulb), whereas the Modified Mercalli Scale measures what the shaking effects were – in other words, what it felt like.

The Richter Scale is not straightforward to comprehend because it is a logarithmic scale, not arithmetic. The difference between a magnitude 5 and a magnitude 6 earthquake is 32 times! And the difference in power escalates rapidly with increase up the scale.

The largest earthquake ever recorded was the 1960 Chilean Earthquake at 9.5. By comparison the Sumatra Earthquake of Boxing Day 2004 was 9.3.

The Alpine Fault

By far the longest single fault in New Zealand, the Alpine Fault, forms a remarkably straight line down the western side of the Southern Alps. It is best seen from space and can be traced for almost 650 kilometres from the entrance of Milford Sound in the south on the Tasman Sea coast of western South Island to the Marlborough coast on the eastern Pacific Ocean side of the South Island.

This, too, is a transpressional fault with significant lateral as well as vertical movement. It last moved in AD 1717 and is considered to move every 100–300 years. It is responsible for the uplift of the Southern Alps that rise to their highest point at Aoraki/Mt Cook, 3754 metres above sea level.

Harold Wellman's Line

Geologists Harold Wellman and Dick Willett are credited with being the first people to recognise the Alpine Fault. Because the West Coast of the South Island is clothed in dense forest, the fault is not at all obvious. Yet with systematic mapping, Wellman was able to demonstrate that for much of its length, the rocks on the eastern side of the fault are schist, whereas rocks on the western side are dominated by granite. He argued that a major fault was the only sensible explanation that could account for the apparent knife-cut juxtaposition of these two very different bodies of rock: schist and granite.

In between, he found a distinctive thin green rock formation that appeared to form a sheet stretching along much of the length of the South Island. Wellman described this rock as the thinnest, straightest and most continuous formation in New Zealand if not the world. It is in fact a zone of intensely crushed rock developed right on the fault itself. It represents the surface of maximum mechanical dislocation, shearing, crushing and movement between two very different blocks of the Earth's crust.

He went on to show convincingly that distinctive rock formations within the Maitai Group of Permian and Triassic age had been offset by more than 460 kilometres.

Within the New Zealand region, the Alpine Fault can be regarded as the cleanest, and most sharply defined segment of the plate boundary between the Pacific Plate to the east and the Australian Plate to the west. It neatly separates continental crust on one plate from continental crust on the other. There are few places on Earth where the public can stop and touch such a conspicuous plate boundary. This is what happens when cream collides with cream: continent versus continent. It is more or less the same type of collision that has given rise to the Himalayas and the European Alps: head-to-head collision of continental crust against continental crust.

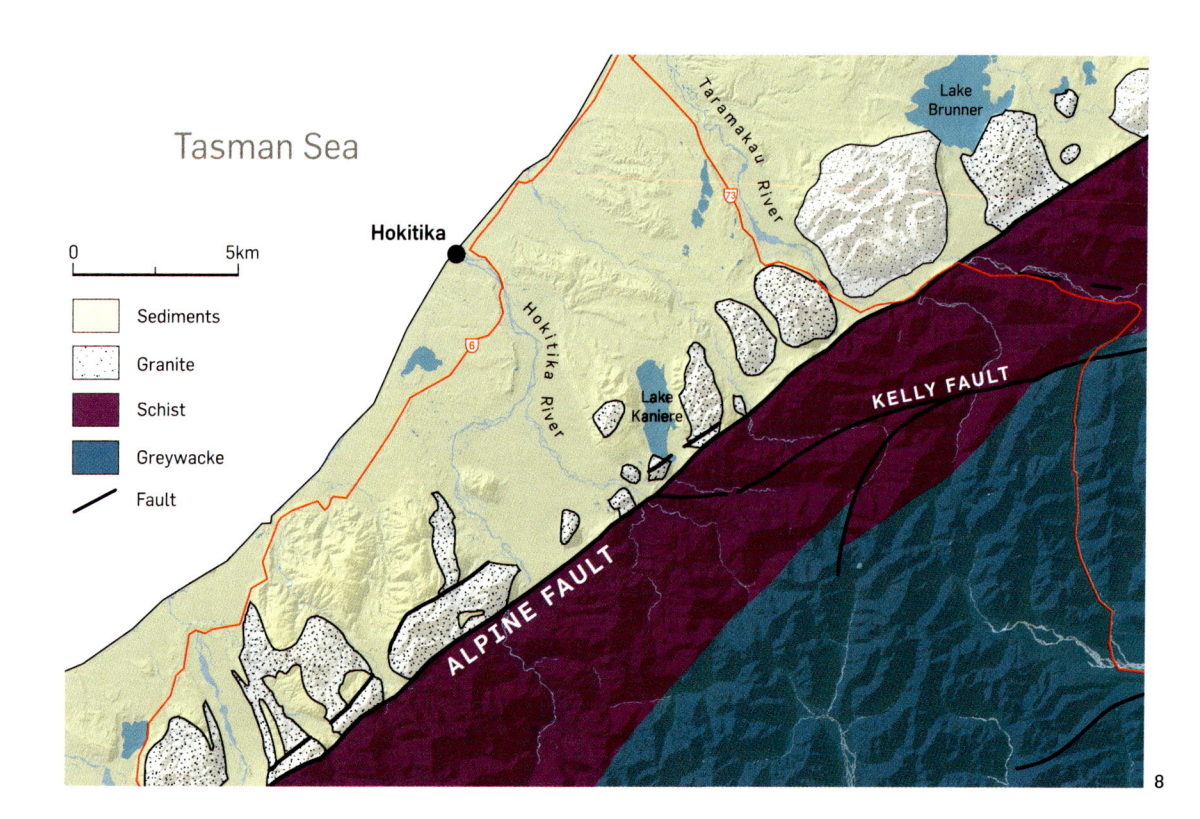

Haremare Creek near Franz Josef is one of the easiest places to see the Alpine Fault and touch the contact. As a rule the fault itself is hard to find. The reason for this is that the rock immediately adjacent to the fault is intensely crushed and inherently weak. So it erodes relatively easily to produce soil that supports lush vegetation. For much of the length of the Alpine Fault, the fault appears as a recessive feature in the landscape, a heavily vegetated slot. No wonder it was hard to find.

Lyell Connects Earthquakes and Faults

The causal relationship between earthquakes and faults is taken for granted these days, but it was established as recently as the 1850s. It was Charles Lyell who made the connection. He is one of the most significant geologists of all time. A Scot, he wrote his famous *Principles of Geology*, and is widely credited with establishing geology as a science.

What is especially remarkable is that Lyell made the connection between earthquakes and faults, based on eyewitness accounts and measurements made in New Zealand, specifically in relation to the 1855 Wairarapa Earthquake. He himself never visited New Zealand.

During March and April 1856, in London, Lyell had the opportunity to question closely three men who had experienced the quake first-hand and who had made excellent observations. One was Edward Roberts, an engineer with the Royal Engineers; another was Walter Mantell, son of the famous paleontologist Gideon Mantell, and an employee of the New Zealand Company; and lastly, Frederick Weld, who established the first sheep station in the South Island at Flaxbourne in Marlborough, and who later became a Premier of New Zealand.

As a result of his interrogation of these eyewitnesses, Lyell presented a number of talks to the scientific establishment on the Wairarapa Fault in both England and France, with wonderful titles such as, 'On the Successive Changes of the Temple of Serapis'.* In time, these lectures were published and they are now regarded as benchmark papers within the history and philosophy of science. Lyell was the first person to demonstrate the link between earthquakes and faults.

Earthquakes and Faults: Chicken and Egg?

Earthquakes are best thought of as huge pulses of energy that are released when large blocks of the Earth's crust snap. The crust is literally 'broken'. It is not very different from breaking a wooden or plastic ruler. It is not correct to say that faults generate earthquakes or that earthquakes generate faults. It is not a chicken-and-egg situation. They are both manifestations of the same thing, the same process. They are inextricably linked.

You can only snap a solid: it is not possible to snap a fluid. This is why earthquakes only occur within the outer part of the

8. On either side of the Alpine Fault the rock types differ markedly. Granite to the west, schist and greywacke to the east – forming the Southern Alps. This map is of the Hokitika area of the South Island.

9. Regarded as the 'father of geology', Charles Lyell wrote the seminal three-volume *Principles of Geology* between 1830 and 1833. Although he never visited New Zealand, he relied on eyewitness accounts of the 1855 Wairarapa Earthquake to make the link between earthquakes and faults.

---

*The curious title of Lyell's talk alludes to ruins of an ancient Roman town at the coastal town of Pozzuoli (near Naples) where a statue of the god Serapis (a Romanised version of the Egyptian God Osiris) was excavated. A drawing of the so-called 'Temple of Serapis' appears as the frontispiece to Lyell's *Principles of Geology*. His talk related to his newly gained wisdom about earthquakes and faults from New Zealand. At last he could satisfactorily explain the amazing history of the Pozzuoli area: the destruction from the shaking of earthquakes, but, more significantly, the drowning of the Temple as a consequence of faulting: it had been down-dropped several metres below sea level. Since then the reverse has happened: faulting has raised the area again by several metres, more or less back to where it was when it was built. All this can be figured out from careful examination of the columns still standing within the Temple. Halfway up the columns there are tell-tale borings made by marine molluscs of the bivalve *Lithophaga*.

Shallow earthquakes (< 40km)

Deep earthquakes

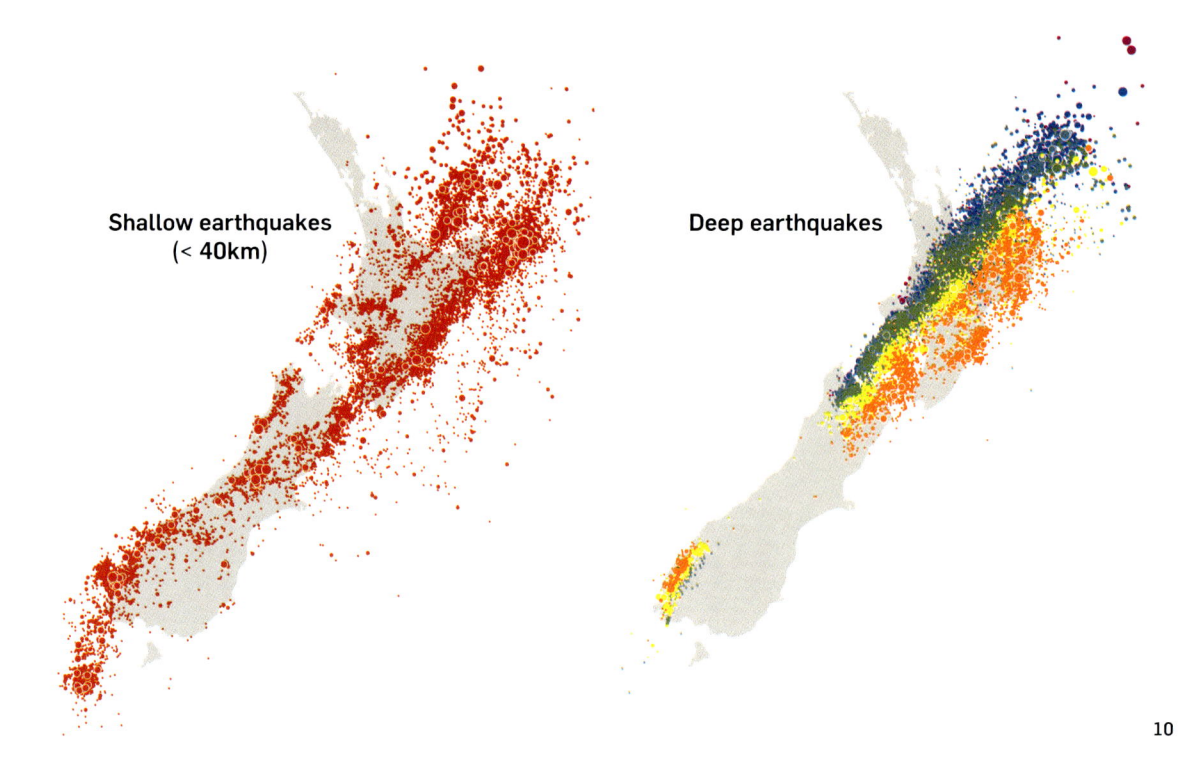

10. When New Zealand earthquakes are plotted on a map, they reveal a distinctive pattern. Shallow earthquakes (less than 40 km deep) occur along the plate boundary, including along the Alpine Fault. Deep earthquakes (more than 40 km deep) also occur along the plate boundary, but curiously not along the Alpine Fault. The shallowest deep earthquakes (orange) are those nearest the subduction zone. Increasing depth is shown by successive yellow, green, blue and purple colours. Note how in the north the deepest quakes occur to the west, while in the south the situation is reversed; the further from the subduction zone, the deeper the tremor. These maps present 10 years of activity (1990-1999) and the bigger the dot the bigger the earthquake.

Earth – the crust and upper mantle. With increasing depth and increasing pressure and temperature conditions, the solid rock that makes up the crust and mantle becomes more fluid. The mantle behaves like a fluid, even though it is neither liquid nor molten.

We know all this from geophysics, the study of the four principal physical properties of the Earth (gravity, magnetics, seismology and electrical properties), as well as experimental simulation work in laboratories specialising in high-pressure and high-temperature conditions.

The Terror of Speed

The worst aspect of earthquakes is their speed. They always come in two very distinct waves. One is faster than the other and yet both are generated at exactly the same moment. The first wave (the principal or 'P' wave) travels at about 20,000 kilometres per hour. This is about five times the speed of a bullet or 20 times that of a jumbo jet. The second wave (the shear or 'S' wave) travels at about half this speed. So, this means that the P wave always arrives before the S wave. The time difference between their arrival times is a measure of the distance to the earthquake focal point, the place of initial rupture. Seismologists call this point the source (or hypocentre), and the point directly above it at the Earth's surface is the epicentre.

P waves and S waves behave quite differently. P waves can transform into sound waves. So, when people hear an earthquake, this is what they are hearing: the P wave hitting the rock–air interface with some of the energy becoming audible sound. S waves are the really damaging waves and, strange but true, they cannot pass through fluid. These various properties have enabled us to work out all sorts of things about the structure of the Earth. Most interestingly, it is the reason for thinking that the Earth's outer core is liquid: S waves cannot pass through it.

The Nature of the Ground

We humans, and indeed all other animals, are anatomically designed to cope

11. One of New Zealand's most recent major earthquakes occurred near Edgecumbe in 1987 when the brittle crust of the land was torn open by rifting of the Taupo Volcanic Zone along the Edgecumbe Fault. Most people assume that the kinks in the railway lines are to do with the fault, but they are not. The kinks pick out the hidden location of gullies in the landscape prior to the railway engineers forming a flat surface and laying down the tracks. The earthquake has exploited these slightly less dense weaker areas of ground and the shaking effects have been dramatically amplified.

with earthquakes. We have inbuilt suspension systems that are very flexible and accommodating. Sudden movements are no big deal for us. However, the same is not true of buildings. Sensible planning and cunning devices are needed to protect buildings from collapse during the intense shaking of earthquakes.

The nature of the ground beneath a building is immensely important, especially the top metre or so. Consider those extraordinary images of the railway track after the Edgecumbe Earthquake: the most impressive zigzags were generated during the quake. Imagine the forces necessary to bend steel railway tracks. How can this have happened? And why are they located in just a few places and not in others? Most people looking at these images think that the kinks in the tracks are located above faults, but this is not the answer.

To Bend Steel

When the railway line was built, a flat surface was formed using a bulldozer and any irregularities in the surface were filled in or scraped off in readiness for laying the sleepers and then the rails. The kinks in the tracks are located above small gullies that were in-filled during this process. The ground in these areas just happens to be less consolidated and therefore less dense than the surrounding un-kinked regions of the track. The difference in density is nevertheless subtle. However, it goes to show the awesome power of nature and its uncanny ability to exploit even the slightest weakness or point of difference. More importantly, it shows just how significant the nature of the ground immediately beneath our buildings is. The top metre is all-important.

The waves of seismic energy coursing up from within the crust are amplified as they pass from dense material to less dense material. The greater the density change, the greater the amplification. Imagine using a stock whip: a simple twitch of your hand translates into an extraordinarily fast-moving oscillation of the tip of the whip, culminating in an audible crack as sound. The energy wave is firstly amplified and then transformed into sound, but most of it is involved in the violent movement of the tip of the whip as the energy attempts to pass from the dense, solid whip to the vacuous air. The tracks above the gullies behaved rather like the tip of a whip and were instantaneously bent.

Those who live in houses on unconsolidated ground, or work in buildings that are built on unconsolidated ground, have reason to be alarmed. An earthquake could destroy such houses and buildings in just a few seconds.

How Do We Know How Often Faults Move?
It is not easy to determine how often a fault moves, but it is worthwhile doing so. This is the easiest way to establish how active a fault is. Does it move every 500 years or every 5000? It is first necessary to find evidence of offset features such as soils, volcanic ash beds, or distinctive rock layers. The next step is to find a way of determining the age of the offset: when did it happen? This kind of enquiry is now a major activity of research geologists in New Zealand and referred to as 'paleoseismology'. They are chasing evidence of past earthquake activity. This sort of information is very useful. It helps us as a society make better-informed decisions about where to build and how to build, and, of course, it is all-important information for setting insurance premiums.

Paleoseismologists like nothing better than digging a large trench at right angles across the line of a fault. Then they can see it in all its glory, measure the offset, determine how it behaves and hopefully find useful materials for dating. The exercise is like a forensic investigation: the trench becomes a crime scene, and the faulted rock face is the dead body. The question is how often has it been faulted and when?

Radiocarbon Dating
A small piece of charcoal or a fossil shell may become key evidence. This may be all that is necessary to establish an age using radiocarbon dating, but the technique will work only on carbon-bearing materials that are less than 50,000 years old. You cannot date anything beyond about 10 half-lives of your chosen isotopic dating method. In the case of carbon 14, the half-life is less than 5000 years. After 10 half-lives, there is so little left of the isotope you are measuring, that it is almost undetectable. So, imagine you have 100 atoms of carbon 14. After 5000 years there would only be 50 atoms; the other 50 have 'decayed' to carbon 12. After 10,000 years, or two half-lives, there are only 25 atoms. After 15,000 years there are less than 13 atoms of carbon 14 left; 20,000 years less than seven; 25,000, less than four; 30,000, less than two; 35,000, less than one atom! Understand?! After 10 half-lives there is almost nothing to detect. The unstable radioactive carbon 14 has all but decayed to carbon 12.

However, there are only so many active faults in New Zealand, about 200 in all, and there will come a time when they have been studied exhaustively, but we have not yet reached that stage.

Modern seismology is paving the way to 'seeing' faults in three dimensions. Analytical techniques have been developed that allow us to use earthquakes to literally 'light up' faults and see what is happening at depth. Perhaps not surprisingly, this research suggests a very strong relationship between fluid release and earthquakes. It turns out that the majority of small earthquakes are caused by hydrofracturing: the breaking of solid rock at depth in response to hydraulic fluid pressure effects.

12. Geologists have identified at least 200 active faults on the New Zealand mainland, many of which are close to population centres. In Wellington there are a number of faults scattered throughout the region. Active faults include those that have moved within the past 100,000 years. Those in red have moved within the past 1000 years.

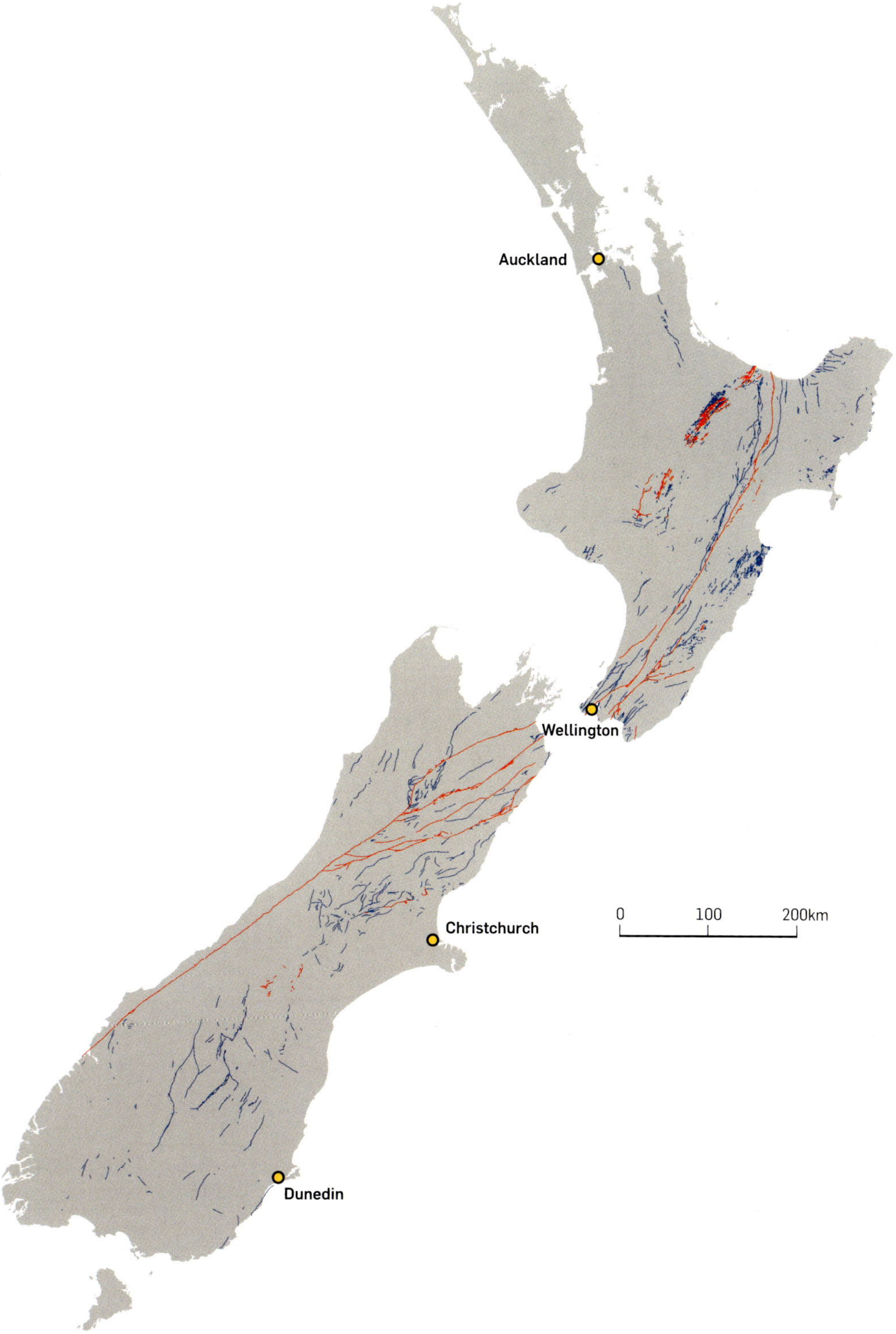

12

11/ Life Aboard Zealandia

Quiet, Stable Zealandia

Following its departure from Gondwanaland, Zealandia was strangely quiet and stable. It was untroubled by plate boundary effects. It behaved like a large, inert blob of cream, riding slowly and sedately across the surface of the Earth in a north-easterly direction. There were no mountain ranges. There was no plate boundary ripping through it. It was subdued and flat, with gentle, rolling topography rather like much of Australia today.

There are two areas of New Zealand where we can see what Zealandia looked like: central Otago and the Chatham Islands.

See Zealandia in the Chathams!

By far the best place is the Chathams. The reason for this is that the islands are so far removed from the

1. The Chatham Islands and central Otago today best resemble what Zealandia would have looked like between 83 and 23 million years ago: a remarkably flat-lying subdued topography. Yet they are not just 'a land uplifted high' as Tasman first saw it, nor a subdued landscape worn down by millions of years of erosion. They are highly modified surfaces, originally land that has subsequently been wave-cut by repeated encroachment (transgression) and retreat (regression) of the sea as the crust has fallen and then risen during the past 83 million years. This view is looking west along the southern coast of Chatham Island. The highest point is just under 300 m above sea level.

2. Paleontologists collecting fossils within Red Bluff Tuff formation, Chatham Island. The fossils are brachiopods, molluscs, echinoderms, corals, bryozoans, sponges and shark teeth of Eocene age. They not only indicate age, but also the nature of the sea floor environment that they lived on as well as the sea itself such as temperature and water depth. These sediments are volcanic ash erupted from a submarine volcano in water depths of several hundred metres.

active modern Pacific Plate boundary, almost 1000 kilometres distant. They appear to have been completely unaffected by the collision tectonics of the past 23 million years.

Northern Chatham Island is stunningly beautiful. Almost devoid of trees, clothed in peat swamp up to 6 metres deep, and punctuated with an array of widely scattered, conical volcanic hills, it is a subdued, low-relief topography. The whole of Zealandia would have looked like this prior to it sinking, although the volcanic hills would have been absent.

There was very little vulcanism in Zealandia. The only place where we can be certain that there was terrestrial vulcanism was in the Chatham Islands, although there is evidence to suggest that perhaps similar shield volcanoes developed elsewhere on the Chatham Rise.

Smaller, intra-plate, hot-spot basaltic volcanoes about the size of those in Auckland erupted during the 30 million years of Zealandia's history from 55–25 million years ago, but they were all submarine. The evidence for this includes small areas of volcanic rocks in the Oamaru–Kakanui area of north Otago, north Canterbury, Westland and the Chatham Islands.

In parts of New Zealand, and in central Otago in particular, there are areas of exhumed Cretaceous (146–65 million years ago) topography, but like the rest of mainland New Zealand it was highly modified by marine abrasion as Zealandia slowly sank. The surface produced by this process is referred to as the Waipounamu Erosion Surface and it is superimposed on the older Cretaceous land surface. Put simply, during the 60-million-year history of Zealandia, the sea had the opportunity to cut and smooth the inherited topography of Gondwanaland. The sea slowly encroached on a sinking Zealandia.

Recent research has demonstrated that the old Cretaceous land surface, the so-called 'peneplain', is not a planar geomorphic feature as originally perceived. This will come as a shock to many New Zealanders who have been taught that extensive flat surfaces in New Zealand's landscape are the product of the slow wearing down of land by erosion to form a peneplain.

However, the geology of the peneplain has been carefully unpacked to reveal a very different story. For generations, we have all been tricked! We now know that where preserved, the old Cretaceous land surface has quite marked relief. It is not planar and is always overlain by fluvial sediments. The conspicuous planar surfaces in our landscape are all marine-cut. This is the Waipounamu Erosion Surface, and where sediments are preserved above it, they are always marine.

As if the interplay between these two surfaces is not complicated enough, they have subsequently been warped and folded by tectonic collisions within the past 23 million years of New Zealand's history. These areas have been stripped of the sediments that accumulated as Zealandia sank. The surface really does look ancient and worn, close enough though to what Zealandia looked like prior to it sinking.

Some depictions of dinosaurs in Zealandia have a backdrop of large active volcanoes. However, there is very little geological evidence of vulcanism in Zealandia. The only significant volcano to erupt during its 60-million-year history (85–25 million years ago) was in the Chathams.

Chatham Volcano: Zealandia's Biggest Volcano

The main southern mass of Chatham Island is the remnant of a large shield volcano that erupted 80–70 million years ago. These rocks are known as the Southern Volcanics. They are comprised of a stack of flat-lying lava flows with red fossil soils between them. The flows were all erupted on land.

Strangely, this volcano is unnamed, despite the fact that it is of great significance to the history of Zealandia. It is best referred to as the Chatham Volcano. It cannot be called Chatham Island Volcano because it was not an island then. The entire Chatham Rise was a great southward-facing elongate promontory of Zealandia.

Chatham Volcano is similar to the Dunedin Volcano that erupted 16–10 million years ago, and the Lyttelton Volcano that erupted 12–6 million years ago, but it is obviously much older and is also more eroded. In some ways, it was more like a Hawaiian volcano, producing a huge, broad, shield-shaped volcano at least 50 kilometres across. Only the northern flank of Chatham Volcano is preserved, and it would have erupted from a centre in Pitt Strait.

Notable outer parts of Chatham Volcano occur beyond southern Chatham Island. These include the lava flows with spectacular basaltic columns in Ohira Bay on northern Chatham Island. This is a major tourist site and must constitute the closest formation New Zealand has to Northern Ireland's Giant's Causeway. On Pitt Island there is the equally spectacular Hakepa Hill, famous for its claim to have been the first place on Earth to witness the sunrise on the first day of the twenty-first century.

The most remarkable thing about Chatham Volcano is that the rocks have not been deformed: they are more or less in exactly the same place as they were when they erupted and cooled. Rocks of similar age in mainland New Zealand are generally very deformed; indeed they have been pushed around by plates colliding. Not so in the Chatham Islands.

Because of this, the Southern Volcanics, the rocks of Chatham Volcano, are of special geological interest. Furthermore, it turns out that there is no other place on land within the entire Pacific Plate where there is such a fine record of the Earth's magnetic field 80–70 million years ago. It is superbly preserved in the iron-oxide minerals in the basalts in the Southern Volcanics.

The Stratigraphic Record of Zealandia

What do we know of the history of Zealandia? In order to answer this, we need to examine the rocks of New Zealand that are 83–23 million years old. They are the source of everything there is to know. We therefore need to look at all rocks of Late Cretaceous to Late Oligocene age (99.6–23.8 million years ago). This includes rocks of Paleogene age, namely the Paleocene, Eocene and Oligocene.

Rock formations of this age range are not especially widespread on mainland New Zealand today. They have been largely eroded away or they have been buried and are therefore out of sight and not conspicuous at the surface. These formations are remnants of much more extensive sheets of sedimentary rock. We know this for certain from geophysical exploration of submerged Zealandia.

In the endless search for hydrocarbons, the oil exploration industry has inadvertently explored a fair bit of Zealandia using seismic surveying techniques. Every now and then an exploration drill-hole is made. What is found in the drill-core provides hard evidence of the history of Zealandia as recorded in the rocks. Over the years, thousands of metres of drill-core has been painstakingly retrieved and examined in great detail. The drill-core is mostly marine, testimony to the largely submarine history of Zealandia. There are abundant fossils preserved within these sediments, mainly of plankton and shells, but, of course, they are primarily a record of life in the Pacific Ocean. They do not preserve much of a record of life on land, or Zealandia, as it was then.

Fossil Spores and Pollen

However, the drill-core does provide some record, in the form of fossil pollen and spores, which tells us at least something about the plant life on Zealandia. Pollen and spores are so small that they are easily transported in the wind. At times of the year, the air is filled with them, as those with pollen allergies know all too well. Offshore winds carry pollen and spores out to sea and they rain out into the ocean and ultimately some of them accumulate in sediments on the sea floor. This is how they end up in marine sedimentary rocks.

Plants produce vast quantities of pollen and spores. The stuff gets everywhere! The only trouble with fossil pollen is that you cannot always be sure just how far it has been transported, and, therefore, where

3. Fingers of lava from extinct volcanoes stretch out into the ocean off Banks Peninsula. Between 12 and 6 million years ago, lava spewed out of Lyttelton and Akaroa volcanoes to create the hilly volcanic landscape that is a stark contrast to the Canterbury Plains beyond. As sea level rose to its present height about 6000 years ago, the present-day harbours were formed. This view is of Lyttleton Harbour with Sumner (lower right) and Lyttleton (centre).

4. Zealandia's sedimentary heritage. This map shows the present-day distribution and extent of Late Cretaceous to earliest Miocene sedimentary rocks that accumulated during the 60 million long years of Zealandia's history, from its departure from Gondwanaland (83 million years ago) and subsequent slow sinking through to its maximum submergence (23 million years ago).

5. New Zealand's oldest fossil flowers. Gathered by paleobotanist Liz Kennedy (GNS Science) from Rakopi Formation rocks exposed at Pakawau, near Farewell Spit, this amazing specimen, one of many, is of Late Cretaceous age. It resembles a mallow, but its true floral affinity is completely unknown. This flower would have bloomed on Zealandia at a time when the Tasman Sea was much smaller. Nevertheless, the Zealandian flora may have already acquired its own character, distinct from that of Gondwanaland.

it has come from. Given the right weather conditions, pollen can be transported thousands of kilometres.

Pollens vary tremendously. Some are delicate and some are very robust; some are large and some are small; some are ornate and some are very plain. This is the world of the palynologist.

GNS Science employs at least four palynologists. This is probably the biggest grouping of such specialist paleontologists in the southern hemisphere. With detailed knowledge and expertise, a great deal of useful information can be gained, especially about the composition of the flora and the environmental conditions at the time.

Zealandian Coal

Fortunately, there is a better record of terrestrial life on Zealandia: in our coal measures. Coal forms as dead plant debris accumulates, primarily in swamps and forests. Coal may, therefore, be thought of as a record of terrestrial life and particular environmental conditions. After all, coal does not just form anywhere.

The older sedimentary rocks of Zealandia are terrestrial or freshwater sediments of Late Cretaceous to Eocene age. They accumulated in rivers, lakes and swamps within lush, wet forest 75–45 million years ago. The fact that the coal deposits are or were extensive is clear evidence of the low-relief, gentle, subdued topography of Zealandia.

Zealandian coal underpins the New Zealand coal industry and is mined or was mined, wherever it is easily accessible. Late Cretaceous coals occur on both the east and west coasts of the South Island, notably near Westport, Greymouth, Ohai and Kaitangata, but also elsewhere. Younger Eocene coals occur widely, especially in the North Island, such as in the Waikato and Northland.

Coal and Fossils?

For all that coal is of organic origin, the original plant material is usually unrecognisable. It has been highly altered and transformed into a heady chemical blend of hydrocarbons. This process destroys fossils so, as a general rule, coal measures are unsuitable rocks for fossil preservation. Acids, in particular sulphuric acid, are produced in the process of making coal, and these dissolve bones and shells. So, for all that coals are indeed indicative of land, the chances of finding fossil bones are remote. Coal mining occurs the world over, but, with some notable exceptions, has produced notoriously few fossil finds.

Amber, 'Jurassic Park' and Kauri Gum

Amber, or fossil resin, is probably the most common

fossil-bearing material within coal, and, as everyone knows from the film *Jurassic Park*, insects are commonly preserved in amber. At least this much is true! The chances of recoverable DNA being preserved in blood ingested by mosquitoes is a nonsense and an impossibility. If you think of an insect such as a fly as essentially a bag of fluid, in this respect it is like any other animal body. If you trap the fly in an impervious liquid glue such as resin, the bacteria within the gut of the dead fly will take over and survive on the fluids trapped within the fly. The bacteria will exhaust the resources available for existence until they too perish. All that might be preserved is a pile of spent bacteria.

Amber is a relatively common material found in Zealandian coals, but as yet has not been systematically collected and studied for preserved insects, or any other fossils for that matter. There are records of small animals, including amphibians, mammals, reptiles and birds, being found in amber, but none from New Zealand as yet.

Kauri gum is resin derived from the New Zealand kauri tree, and gum-digging was a significant industry in the northern North Island in the late 1800s and early 1900s. The gum was extracted from shallow but nevertheless buried forest floors, flood plains, soils and swamps. They do contain records of fossil insects. However, none are considered to be older than middle Pleistocene, so they are all less than 1 million years old.

Fossil Flowers of Pakawau, Zealandia

Botanical studies of plant fossils from sediments associated with the coal have provided much insight into the nature of the forest plants that grew on Zealandia at this time, particularly the study of fossil pollen and spores. These microscopic fossils are much more abundant, more diverse, and much more easily prepared and examined than fossil leaves, wood, flowers, fruits and seeds.

However, some amazing discoveries have been made, in particular of fossil flowers. The oldest fossil flowers known from New Zealand are Zealandian and, although their identity is as yet uncertain, they resemble flowers of the mallow family. These have been collected by paleobotanist Liz Kennedy. They are from rocks that are about 70 million years old in the Rakopi Formation near Pakawau, in the north-westernmost South Island.

The Zealandian Flora

The coal is dominated by podocarps and araucarians. Pollen and spore fossils are preserved from ferns, bryophytes, horsetails, lycopods and flowering plants (angiosperms). Tree ferns were abundant in some areas as determined on the basis of abundant spores. Both macrofossils and microfossils of lycopods, horsetails, ground ferns and hepatics are known.

The forests of Zealandia included abundant and diverse flowering plants, especially of protea and southern beech trees, along with ancestors of modern flowering plants such as pinks, wallflowers, heaths, grass trees, hollies, lilies, mistletoes, pepper trees, buttercups, polypodiums, celery pines and totara.

Zealandian plant fossils of Eocene age are preserved in some areas of New Zealand, including leaf fossils and some spectacular spiny seeds from north Otago and the Dunedin area.

This flora was inherited from Gondwanaland, but then would have evolved over the 60 million years of Zealandian history. Initially Zealandia was a vast land mass almost half the size of Australia, but with time it became increasingly more remote from Gondwanaland and shrank in area as it slowly sank.

The Zealandian flora was very different from the modern flora of New Zealand. It is likely that all endemic species of the modern New Zealand flora have evolved on New Zealand within the past 23 million years, not on Zealandia. Furthermore,

6. A fossil spiny seed of *Casuarina*. This was collected from lake sediments of Miocene age preserved near Bannockburn, central Otago. It is testimony to warm, arid conditions during at least some of Miocene time, and the existence of a sub-tropical flora not unlike Australia today.

7. An Eocene marchfly larva, *Dilophus campbelli*, (final instar larval exuviae). This is one of very few fossil insects recorded from New Zealand, and the first pre-Pleistocene insect described. A: The actual fossil, almost 20 mm long, preserved in siltstone and collected from a locality near Livingstone, inland of Oamaru.

B: A sketch of what the fossil reveals to an entomologist such as Tony Harris who described and identified it. This marchfly lived on Zealandia more than 50 million years ago. Amazingly, it closely resembles modern day *Dilophus nigrostigma*, New Zealand's largest species of marchfly.

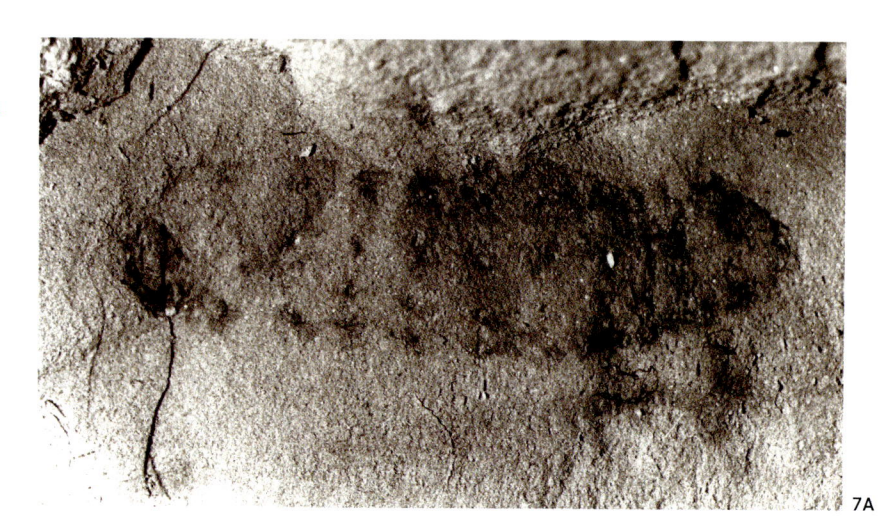

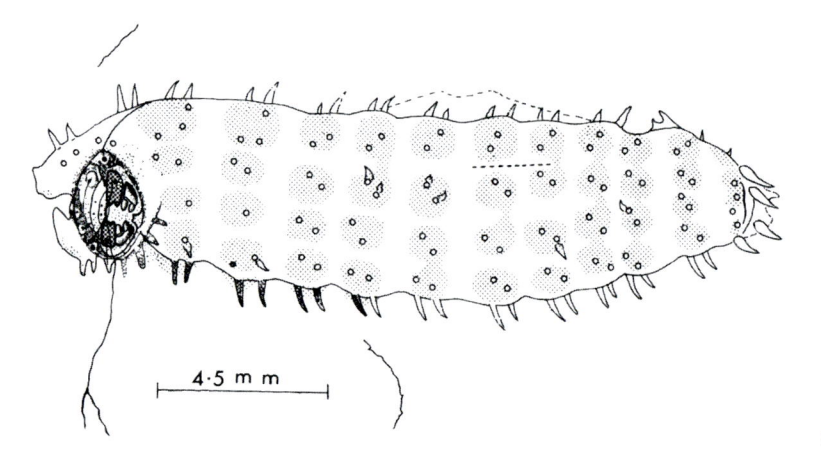

although there would have been relict Gondwanan floras on Zealandia, there are none within the modern-day flora of New Zealand.

By the time of maximum immersion of Zealandia 23 million years ago, the only record of the flora is that of fossil spores and pollen, and it is very meagre. Notably, it is also very different from that of the earliest fossil floras recorded from emerging New Zealand in the Early Miocene age. Although it cannot be proved geologically one way or the other, it is conceivable that Zealandia was entirely submerged 23 million years ago. At most, there may have been a few small islands.

Zealandian Islands: Where Might They Have Been?

The only bit of what is now the North Island that was upstanding enough and that may have been an island 23 million years ago is the Herangi Range north of Awakino and south of Kawhia Harbour on the west coast. There may also have been small islands to the north of Northland in what is now the Norfolk Basin. The evidence for these islands is based on the existence of remnant flat surfaces, presumably wave-cut, at considerable depths, in excess of 1000 metres. Dredging of these surfaces has recovered fossils of shallow-water organisms. So, it is most likely that these areas were once at or above sea level and have subsequently sunk. In the South Island, the most plausible location of islands 23 million years ago is within Fiordland. One other area that might have had islands is the Catlins region.

Previous attempts by geologists to depict the extent of land in New Zealand during Late Oligocene to Early Miocene time all show islands in these areas. However, modern analysis of the geological evidence for the existence of these putative islands has raised the probability that the entire region of modern-day New Zealand was totally submerged.

What About the Zealandian Fauna?

Very little is known about the animals of Zealandia. The only terrestrial animal fossils known are the pterosaurs, dinosaurs and turtles of inland Hawke's Bay (75 million years old). No other reptilian fossils have been recorded, and no mammalian, bird or amphibian fossils, bearing in mind that by 'Zealandian' we are referring to that period of geological time 83–23 million years ago.

Not a single Zealandian fossil is known of any ancestors of the modern terrestrial reptile, bird and mammal fauna of New Zealand. However, absence of terrestrial fossils does not necessarily mean absence of life. Far from it!

Zealandian Insects

Rare insect fossils are known. Fragmentary remains are routinely recovered during sample preparation for palynology, but there has been no systematic study. Trace fossils that may be interpreted as insect borings have been observed in fossil wood.

By far the most significant Zealandian fossil insect is that of a final instar larval exuviae of the mayfly family Bibionidae. The fossil is of the discarded skin of *Dilophus campbelli*, which appears to be almost indistinguishable from the common extant marchfly *Dilophus nigrostigma*! The fossil resembles a small caterpillar and was found in lake sediments near Livingston, inland from Oamaru. It was in very finely laminated mudstone with abundant leaf fossils and also shells of the freshwater clam *Hydridella*. The age is uncertain but it is thought to be Early Eocene, 56–49 million years old.

Claiming Charles Fleming's Bounty!

There is quite a story to this find. It was collected by one of the authors (Hamish Campbell) while on a collecting trip with three other paleontologists (Doug Campbell, Ian Raine, Jonathan Aitcheson) and was subsequently identified by Dunedin-based entomologist Tony Harris. At the time, the famous New Zealand paleontologist Sir Charles Fleming had a bounty out for the first pre-Pleistocene fossil insect (that is, one more than 1.8 million years old). He was contacted and once he was satisfied that the claim was authentic, the handsome prize was handed over: 15 pounds sterling drawn on a British bank. As the senior member of the collecting party, Doug Campbell was sent the cheque. However, rather than cash it in, he chose to cut the cheque up neatly into four pieces!

Zealandian Seas

Although the terrestrial record of Zealandia is scant, the marine record is good. Late Cretaceous, Paleocene and Eocene marine sediments provide a wonderful insight into life in the south-western Pacific and Tasman Sea before the last big event in the break-up of Gondwanaland – the separation of Australia and Antarctica. These records are significant because oceanic circulation was so different then, as was climate. There was no such thing as the Circum-Antarctic Current, nor were there polar ice caps. It was an utterly different world from today. Antarctica only established its polar position 34–33 million years ago, at the very end of Eocene time, and from this moment onwards the polar ice cap became established and started exerting its huge influence on global climate, which continues to this day.

No wonder New Zealand has attracted so much paleontological interest. We have a fossil record of life in the south-western Pacific since Late Cretaceous time 83 million years ago that is simply unrivalled. It tells the story of the steady, sustained sinking of Zealandia with ever-deepening marine conditions.

8. Charles Fleming was Chief Paleontologist at the New Zealand Geological Survey. He was a tireless natural scientist with broad interests and enthusiasms, not least paleontology, environmental conservation and the Royal Society of New Zealand. He was especially interested in the origins and antiquity of New Zealand's fauna and flora. Part of his legacy is the National Paleontology Collection and in particular a major collection of comparative fossils from overseas.

9. Tony Harris (Otago Museum) is a well-known entomologist and science communicator.

10. Trace fossils (ichnofossils) are surprisingly common and their study (ichnology) is a science all on its own. Trace fossils are present in most sediments and sedimentary rocks. They may be thought of as the traces of animal and plant behaviour. They are records of movement, burrowing, feeding, defacation, struggle and rest. This pebble from Ward Beach in Marlborough is riddled with trace fossils made by a variety of small organisms that lived in the sediment just beneath the sea floor during Paleocene time.

11. *Waiparaconus zelandicus*, a strange fossil that resembles asparagus in terms of not only its shape but also its ornamentation. The nature of the animal that grew this structure remains enigmatic but it has long been thought of as a highly specialised barnacle. This was collected in greensand of Paleocene age from the Waipara River, north Canterbury.

12. The skeletal remains of a large fossil penguin found in limestone of Oligocene age from near Kawhia, west of Hamilton. This spectacular find was discovered by the Hamilton Junior Naturalists Club in early 2006. More than 20 species of penguin are known from New Zealand and the oldest is more than 60 million years old (Paleocene) from northern Canterbury.

Late Cretaceous Marine Fossils

Late Cretaceous marine fossils that relate to Zealandia and the Paleo-Pacific Ocean include foraminifera, radiolarians, sponges, bryozoans, echinoderms, bivalves, ammonites, belemnites, gastropods and rare brachiopods, crustaceans, fish fossils, shark teeth and bone fossils. Perhaps most interesting of all are rare fragmentary fossils of marine reptiles, such as mosasaurs and elasmosaurs. The most spectacular example is an almost complete skeleton of an elasmosaur collected from Shag Point north of Dunedin, and on display in Otago Museum.

Rocks with marine fossils of Late Cretaceous age are not so common but do occur in the Dunedin area, north Otago, north Canterbury, Marlborough, Wairarapa, Hawke's Bay, Northland and the Chatham Islands. They can be most easily observed in the Waianakarua Stream south of Oamaru in north Otago, and at Flowerpot Harbour on Pitt Island in the Chatham Islands.

Paleocene Marine Fossils

There are shallow-water sediments of Paleocene age, 65–56 million years ago, but they are restricted to areas within coastal Otago. The best-known occurrence is along the Wangaloa coast to the north of Kaitangata. Here there are sandstones rich in elongate tapering turritellid shells or pencil shells. Other molluscs are also preserved.

Sedimentary rocks of Paleocene age occur elsewhere, especially in Canterbury, Marlborough and the eastern North Island, but are generally of deeper water origin and, although lacking in shelly fossils, they are rich in microfossils such as foraminifera, radiolarians and dinoflagellates.

Trace fossils of Paleocene age are known from many localities, but in particular Te Kaukau Point at the very south-east tip of the North Island, Tioriori on the north coast of Chatham Island, and some beaches in Marlborough. Ward Beach is a wonderful place to visit, with a magnificent pebble beach. Millions of pebbles here are derived from local exposures of marine sedimentary rocks. The pebbles are smooth and many have trace fossils. They include the burrows, traces and trackways of myriads of organisms that must have lived in the original sediment and foraged for food. Most of these organisms would have been crustaceans, echinoderms and worms that lived within the top metre or so immediately below the sea floor.

Perhaps of greatest interest are penguin fossil bones, from Paleocene rocks near Dunedin, in north Canterbury, and also in the Chatham Islands. These are without doubt the oldest 'penguin' fossils globally.

One of the strangest fossils from New Zealand is the bizarre *Waiparaconus*. Although first described from New Zealand, it is also known from Australia and New Caledonia. It looks like petrified asparagus – that is, turned to stone! Originally thought to be a strange barnacle, there are some doubts. But if not a barnacle, then what could it be? What sort of organism produced it? This is just one instance of a fossil that remains mysterious. We are as yet uncertain of its true identity. There are plenty of such mysteries.

Eocene Marine Fossils

Zealandian marine sediments of Eocene age are much more widespread than those of Paleocene, and there is a much greater diversity of interesting fossils, including coconuts, marine turtles, the oldest whale fossils from New Zealand, and a spectacular fossil fish from Pitt Island in the Chathams. There is even a small fossil egg! It is probably a turtle egg, but what it really is remains unclear.

Oamaru Diatomite

Fossil plankton, particularly foraminifera and dinoflagellates, are of huge interest and of commercial significance in the search for oil and gas exploration; they help explorers find their way in the dark, so to speak. One of the more unusual occurrences of plankton can be found near Oamaru. It is a diatomite deposit some 45 metres thick, almost completely comprised of the skeletal remains of single-celled marine algae called diatoms. The deposit has been exploited commercially, as diatomite has unique filtering properties, but the amazing diversity of diatoms has attracted international interest from diatom experts.

Diatoms are a significant component of marine and freshwater ecosystems and form a major component of the very base of the food chain. As with all plants, they are photosynthesising, but form exquisite, shapely, box-like structures from silica. In a way, they are tiny floating glasshouses or hothouses. These tiny plants are stunning architects, having produced amazing homes for themselves.

In the 2007 ANDRILL project in Antarctica, a total of more than 200 metres of diatomite was drilled through. Diatoms feature significantly in ocean life and health, and when they 'bloom' they form extensive deposits on the sea floor, referred to as 'diatom ooze'. Being comprised of silica, they are rock-forming and the resultant rock is diatomite.

One other diatomite referred to in this book is the Middlemarch Maar diatomite. This differs from the Oamaru Diatomite because it is of freshwater origin and is of Early Miocene age. Diatomite is not exactly rare, but it is not especially common either.

13. Phillip Maxwell (New Zealand Geological Survey), one of New Zealand's most able natural scientists and a molluscan paleontologist who worked almost exclusively on the Late Cretaceous–Miocene marine record of Zealandia. Here he is holding up a block of Eocene siltstone with several superbly preserved specimens of turrid gastropod from the Waihao River, near Waimate, south Canterbury.

14. Dieffenbach's locality, on the north coast of Chatham Island. This is of historic significance because it was lost to science for over a century, buried by sand dunes. Yet this fossil locality was the very first to be recorded from New Zealand in the scientific literature (in 1842). It is an Eocene greensand with lenses of limestone and relatively common fossils of an unusual oyster named *Pycnodonte (Notostrea) tarda*.

Phillip Maxwell

Phillip Maxwell made a career from studying the Eocene molluscan fossils of New Zealand, or rather Zealandia. In so doing, he collected numerous other fossils and should therefore be counted as one of New Zealand's foremost fossil collectors. He made an outstanding contribution to science because of his thorough grounding in mathematics and ecological and evolutionary theory. He is certainly one of New Zealand's finest molluscan paleontologists. One of his innovations was the development and clever use of extraction techniques for freeing large numbers of micro-molluscs from rock. He was then able to apply scanning electron microscopy and make modern taxonomic sense of them by studying the very earliest formed parts of the shell structure.

The most extensive areas of Eocene marine sediments are to be found in north Otago, south Canterbury, Northland and the Chatham Islands.

Dieffenbach's Locality

The first fossil locality to be recorded formally from New Zealand was collected and documented by Ernst Dieffenbach, naturalist to the New Zealand Company. He visited the Chatham Islands in May–July 1840 on the vessel *Cuba*, and must be considered as the first geologist to visit the Chathams. His locality is on the north coast of Chatham Island, within rocks of Early Eocene age, and the fossils were oysters. He returned to Britain in 1842 with his collections and submitted them to the British Museum (Natural History). The fossils were identified by John Gray, a curator who published his findings later that same year. Dieffenbach was not the first to find fossils in New Zealand, but he was the first to document the locality accurately and in such a way that scientists could subsequently re-find the locality precisely, and recollect the fossil.

Oddly enough, between the 1920s and the 1980s subsequent visiting geologists were unable to locate Dieffenbach's locality. But it was rediscovered in 1997. The locality had been buried by sand dunes for more than 100 years!

Shark Teeth at Blind Jim's in the Chathams

Eocene (55.5–33.7 million years ago) limestones are especially well represented in the Chatham Islands. There are no limestones of comparable age on mainland New Zealand, so the Chathams have attracted a number of specialist paleontologists researching the fossils that make up the limestone, particularly bryozoans, echinoderms, foraminifera, barnacles and shark teeth. There are several beaches on the western side of Te Whanga Lagoon in central Chatham Island, and in particular at Blind Jim's Creek, where fossil shark teeth abound and are sought after. Tens of thousands of teeth have been collected over the years. They are all derived from erosion of Eocene limestone and are about 50 million years old. Most of the teeth can be attributed to five or so species of shark comparable to the sand sharks and blue sharks of today, but occasionally other oddities are found, including teeth of the huge *Carcharodon*.

Moeraki Boulders

The Eocene rocks of greatest fame in New Zealand are without doubt the Moeraki Boulders that are exposed on Katiki Beach and Moeraki Beach on the Otago coast between Dunedin and Oamaru. These boulders are eroding out of a widely distributed marine mudstone referred to as the Abbotsford Formation, named after a suburb of Dunedin. The boulders are almost perfect spheres and although most are less than 1 metre in diameter, some are up to 2 metres.

Some would have us believe that these perfectly shaped objects are man-made and relate to a mighty exploration fleet of Chinese junks, but this is pure fantasy to say the least, no matter how interesting it sounds. They are completely natural and to geologists they are 'concretions'. As the name suggests, they are concrete-like. They are much harder and tougher than the mudstone in which they have grown, and hence they erode out of the sea cliff and gather on the beach like so many lost balls.

It is true that they have literally grown! They are secondary features, which in this case means that the mudstone accumulated first and then the concretion formed. After being buried deep under piles of sediment, the conditions were just right for the systematic radial and concentric growth of carbonate cement. As a general rule, this process starts with some organic fragment, a piece of shell or wood, which serves as a nucleation point or seed. Once started, it seems the process just runs until a natural limit is reached. The availability of carbonate may be a constraining factor, as may be temperature, pressure, fluid flow characteristics within the mud, or other physical and/or chemical factors.

The net result is a solid sphere that has formed under a specific pressure regime. What happens when something under high pressure is placed in a low-pressure setting is that, unless it is contained, it will either relieve itself slowly, or fail catastrophically and explode. This is what has happened to many of the Moeraki Boulders. With subsequent uplift and erosion, they have gone from a high-pressure setting to a low-pressure setting. As a result, some have exploded or cracked open within the encasing mudstone, and subsequently the cracks have

14

15

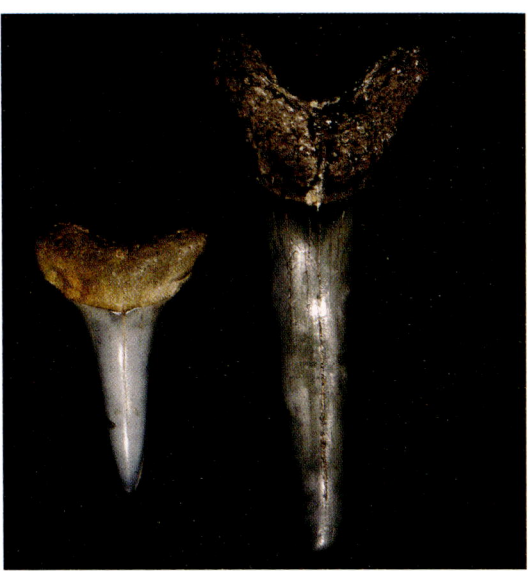

15. Blind Jim's Creek on the west shore of Te Whanga Lagoon on Chatham Island is a well-known tourist spot. Tens of thousands of fossil shark teeth have been collected from here over many decades. They are of Eocene age, derived from the soft limestone cliffs that form the lagoon edge. The waters of the lagoon are responsible for mining the teeth from the limestone and concentrating them on this particular beach. These teeth belonged to a sand shark *Striatolamia macrota*, and are by far the most abundant. The bluish dark colour of the fossils is typical.

16

been in-filled with another, later cement. You could say that the Moeraki Boulders were formed beneath the sea floor of Zealandia, only to be uplifted and fractured with the birth and subsequent growth of New Zealand.

Concretions such as the Moeraki Boulders are surprisingly common and can be observed in numerous mudstone formations throughout New Zealand, particularly on the east coast of the North Island. Some can be observed in road cuts on State Highway 1 between Bulls and Waiouru. However, none are quite as stunning as those at Moeraki, where lines of them can be seen eroding out of cliff faces.

Eocene Slick

The Abbotsford Formation is notorious from an engineering perspective. It is so weak that it fails on slopes of greater than 7°, either natural or man-made. It is widespread in the greater Dunedin area and is responsible for much landsliding and difficult topography. This same formation played a significant role in the Abbotsford Landslide of 1979, not to mention the difficulties of road construction in and out of Dunedin and the necessary removal of Seacliff Hospital.

Geologists have always recognised the problems, but road-planning authorities and engineers do not always take heed. There are comparable mudstones in many parts of New Zealand, but none are quite so treacherous as the Abbotsford Formation. The reason for this has to do with its unique clay mineralogy, the particular mix and relative proportions of swelling clays.

Zealandian Oil and Gas: Crash Energy

As Zealandia sank, the older, organic-rich terrestrial sediments were buried beneath increasingly younger sediments and rocks. This 'loading' is an essential part in the natural manufacture of those popular sources of easy energy: oil and gas. As with cooking, the right pressure and temperature conditions have to be achieved. As a general rule, sediments have to be buried by 5 kilometres, and held there for some time. This is no small requirement! In order to generate oil and gas, you have to bury organic-rich sediments such as the Zealandian coals of Late Cretaceous to Eocene age beneath 5000 metres of younger sediments and rocks. And yet this is exactly what has happened in New Zealand.

Today in Taranaki we are exploiting hydrocarbons (oil and gas) that are derived from Zealandian coal measures. They are the source rocks. However, they were only buried by the requisite thickness of sediments because of the birth and uplift of New Zealand. The hydrocarbon exploration and production industry in New Zealand owes its existence and livelihood to plate collision, the big crash that is ongoing and has been vigorous since earliest Miocene time.

Usually, the fluids (oil and gas) reside in younger rocks that overlie the source rocks. This is because fluids generated in the 'cooking' process are mobile and will always move from a high-pressure to a low-pressure environment. Within the crust, this means upwards. With any luck, they become trapped and can subsequently be found and used as fuel. In Taranaki, a favourite host rock or reservoir rock is fractured limestone of Oligocene to earliest Miocene age. But younger host rocks of Miocene and even Pliocene age are also worthy of investigation.

Conclusion

In this chapter, we have considered life aboard Zealandia during late Cretaceous to Eocene time. The record of terrestrial life is surprisingly meagre, save a reasonable pollen record. The marine record, on the other hand, is considerable, so at the very least, we have an excellent record of the seas that bathed Zealandia.

The next chapter explores the maximum sinking and immersion history of Zealandia during Oligocene and earliest Miocene time.

16. The Moeraki Boulders are concretions or spheres of naturally concreted silt. Here they can be seen bathing on the north Otago coast at Moeraki, like so many discarded beach balls. They are much tougher than the Eocene marine siltstones that they have been eroded from, and hence they accumulate at the base of the cliff, released by the sea.

12/ The Immersion of Zealandia

Set Adrift, Only to Sink

As Zealandia commenced its ship-like voyage from Gondwanaland, it began to sink. We know this from the geological record preserved on land in New Zealand, and from the detailed exploration of what lies beneath the sea floor of submarine Zealandia.

The slow sinking of Zealandia lasted for 60 million years, something we understand from our knowledge of the nature and distribution of sedimentary rocks 83–23 million years old, representing Late Cretaceous to earliest Miocene time. We have considered aspects of several periods in the history of Zealandia: Late Cretaceous (83–65 million years ago); Paleocene (65–55.5 million years ago); and Eocene (55.5–33.7 million years ago). Here we consider the 10 million years of Oligocene history from about 34–24 million years ago. It was during this time period that Zealandia approached maximum inundation. We address the evidence for this phase of Zealandia's history, and consider just how far it sank.

The Nature of Sediments

The nature of the original sediments that make up these rocks enables us to determine the character of the greater environment in which they formed: firstly, where the sediments were derived from (in other words, what the source rocks were); secondly, what water depth they were deposited in; and thirdly, what distance they were from land. These are the three key parameters that enable us to determine the approximate location of the line of zero water depth (that is, the position of the ancient shoreline) through geological time.

In order to determine the nature of the sediments and the rocks they form, they have to be visited in 'the field', recorded and sampled. To many geologists, this is the best part of their job: getting outdoors and doing original research.

The samples then have to be analysed in laboratories by various specialists. The preparation of a rock sample for microscopic examination is a specialist process in itself, normally the preserve of the 'thin section laboratory'. Here, rocks are cut and polished, and then a very thin wafer of rock is prepared, so thin (0.03 millimetres thick) that light can pass through it to reveal its mineral composition. The rock is now ready for microscopic examination using a petrographic microscope (a microscope for examining the light transmission and absorption properties of crystalline materials such as minerals).

It is rather like forensic work. Think of the sediment or rock in the field as a crime scene. The samples will help determine what the sediment is, how it got there, how old it is, where it came from and what has happened to it. The laboratory is used to take the rock to bits and reconstruct how it formed.

Some rocks are crushed so that selected minerals can be analysed in various ways. Geologists commonly use X-ray techniques to identify the less conspicuous minerals present in the rock, especially clay minerals. The relative proportions of minerals can be meaningful, in terms of both mineral type and abundance. In order to determine abundance, statistical counting and measuring techniques are used.

The Fossil Stuff

So far, only the mineral component of the sediments has been discussed. All sediments, and hence all sedimentary rocks, are comprised of material derived from pre-existing rocks and fossil material. The fossil material includes biogenic minerals derived from the hard parts of plants and animals (shells, bone, teeth), as well as tough, non-mineral organic materials (wood, pollen, spores), and fossil organic molecules derived from all life (especially the most common form of life: bacteria).

There probably is no such thing in nature as an organic-free rock or mineral surface, especially in an aqueous environment. The minute a 'fresh' rock surface is exposed, it attracts organic molecules that are ubiquitous; they are everywhere, but are so small that they cannot be seen without the aid of microscopes. Here on the surface of our planet, the inorganic and organic worlds are inextricably mixed.

To study the fossils preserved in a sedimentary rock, clever extraction techniques are used. The technique used depends on what fossils are being studied. For fossil pollen and spores, it is common practice to dissolve the rock in powerful acid. For fossil shells and bone, the easiest method is to use a percussion instrument, rather like a dentist's drill or a small jackhammer. Once the fossils are free of the rock matrix, they can be identified and interpreted.

If fossils occur in large numbers, as microfossils do, statistical studies can be made that may shed light on aspects of the environmental conditions in which the original microscopic plants and animals thrived – for example, the availability of nutrients, or the temperature, depth and salinity of the sea water. Such studies provide a measure of productivity and diversity. All this information can help determine where in the ocean the sediments accumulated, which in turn is hugely significant in making sense of the submergence history of Zealandia.

Mapping Sedimentary Rock Formations

Our knowledge of the 'distribution' of Late Cretaceous to Miocene sedimentary rocks is based on geological mapping.

In 2007 GNS Science was completing a major project known as QMap (an acronym for 'quarter million'), which commenced in the mid-1990s. It presents the geology of the whole country in 21 map sheets. It is a completely new 1:250,000 geological compilation of the whole of New Zealand. It is 'new' because it had previously been mapped at 1:250,000 (in the 1960s). However, the 'old' map just predated the advent of plate tectonics, which has completely revolutionised our understanding of geology. Although the distribution of rock formations has changed little, their interpretation has changed significantly.

At this scale, a distance of 10 kilometres is represented on the map by just 2.5 centimetres. Nevertheless, it is the most accurate and up-to-date geological map of New Zealand and it shows the distribution of Late Cretaceous to Miocene rocks beautifully. We know exactly where they are! It should be noted that significant parts of New Zealand are also mapped at other scales, such as 1:50,000, and some at 1:25,000, especially urban areas and areas of economic importance.

1. Understanding ancient New Zealand depends heavily on geological maps that reveal the distribution of rock formations in terms of their composition (lithology) and age. This group from GNS Science includes many of those involved in the preparation of 1:250,000 maps of New Zealand's geology, affectionately known as QMap. From left to right: Dougal Townsend, Steve Edbrooke, Mo Turnbull, Julie Lee, Mike Isaac, John Begg, Mark Rattenbury. Looking over their shoulders is Alexander McKay, one of New Zealand's great early geologists.

2. James Hector, first Director of the New Zealand Geological Survey, here dressed in his robes as Chancellor of the University of New Zealand. For decades he was the face of science in New Zealand, and the most trusted advisor on matters of science to the New Zealand Government.

The First Geological Maps of New Zealand

In 1865, with the establishment of Wellington as the capital of New Zealand, James Hector was appointed as Director of the New Zealand Geological Survey. He and his staff set about mapping the geology of New Zealand. The national interest was paramount, and in particular the search for natural wealth: gold, silver, copper, coal and other valuable minerals that might be found.

However, prior to 1876 the country was organised in terms of provinces, and most provincial governments employed a geologist who was charged with surveying duties. Hector had originally come to New Zealand as the first 'provincial geologist' for Otago, and was succeeded by Frederick Wollaston Hutton. Ferdinand von Hochstetter was hired by Nelson province, and Julius von Haast became the provincial geologist for Canterbury. James Coutts Crawford was Wellington's token geologist, but he was the man who successfully influenced the New Zealand Government, and in particular none other than Frederick Weld (who reported on the effects of the 1855 Wairarapa Earthquake to Charles Lyell, of the need for a national geological survey).

The names of these famous men are all associated with the earliest known geological maps in New Zealand. With the rise of universities and the establishment of 'geology departments', geological mapping became a matter of academic training, professors and research students.

With all this sustained effort over a period of almost 150 years, we can say with great certainty that the distribution of Late Cretaceous to Miocene sedimentary rocks in New Zealand is well known and well understood. However, it is the interpretation of the rocks that continues to exercise the minds of geologists. What meaning do the rocks provide? What more can they tell us? What undiscovered treasures do they contain?

The Zealandian Cover Sequence

What the map shows is that in places there is a recurring pattern, a motif. These are the remnants of that orderly sequence or succession of layers of sedimentary rock that was systematically draped, sheet-like, over Zealandia. This was a prolonged process, spanning several geological periods, 83–23 million years ago. Much has

3. As if the ground has been drilled, this extraordinary landscape inland from Timaru is classic karst or sinkhole topography developed on limestone. This type of erosion, due to dissolution from rainwater, is surprisingly common in limestone country throughout New Zealand and indeed globally. The reason is that limestone is largely comprised of a single mineral, namely calcite, that is easily dissolved in the presence of acid, even though it may be exceptionally weak.

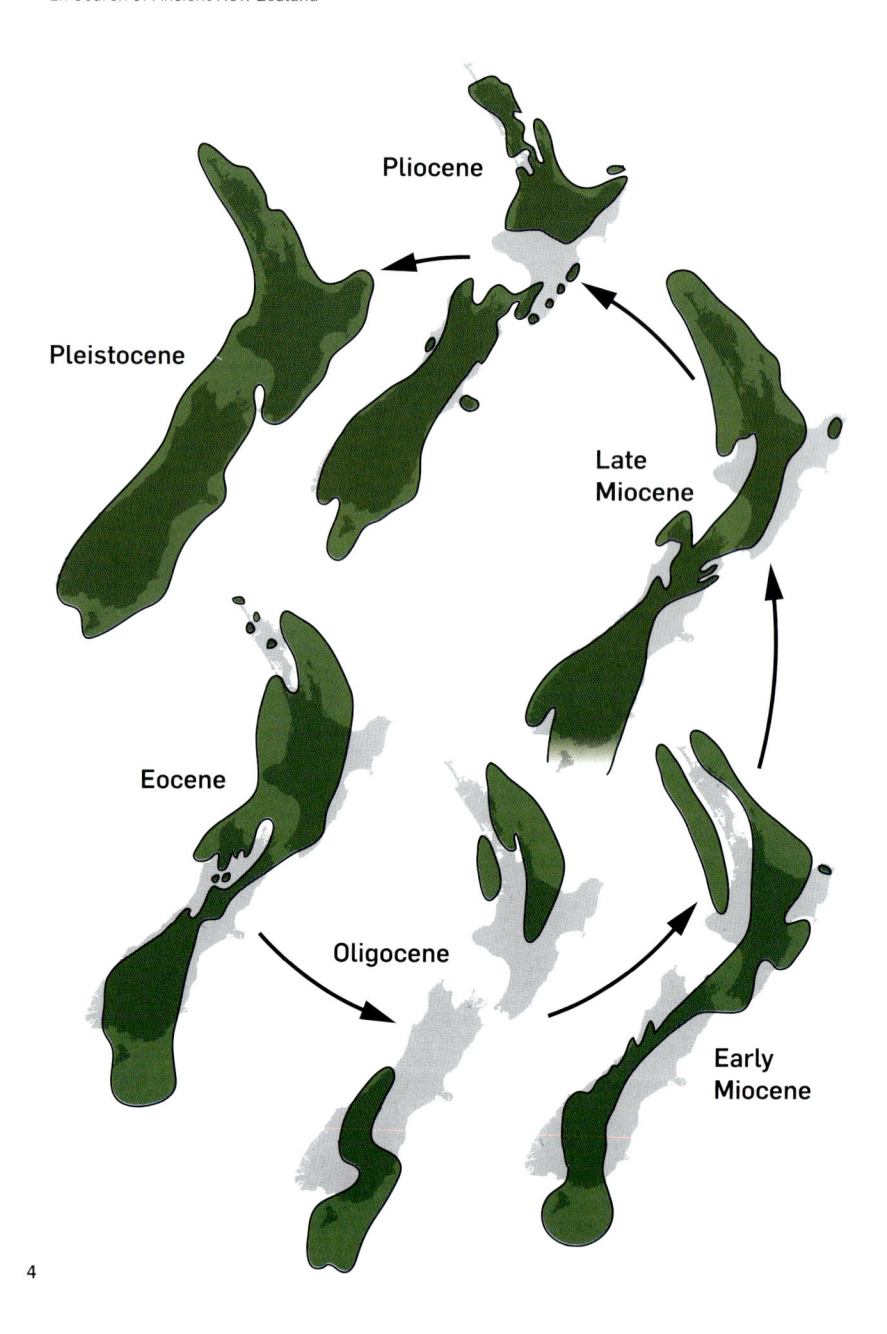

changed since then, so this orderly sequence is preserved only in certain places and no longer forms a continuous cover. Perhaps it is best thought of as a quilted cover, with many of the segments of quilt gone, or appearing to have gone. The reasons for this are that since earliest Miocene time, that is within the past 23 million years, the cover has either been buried by tectonic down-drop or completely eroded away by tectonic uplift.

This sequence of sediments records the slow flooding of Zealandia by the sea. To put it another way, the sea transgressed upon the land, and in so doing a distinctive pattern of sediment distribution was produced on a regional scale. This is known to geologists as a 'transgressive marine sequence'.

The base of the sequence is characterised by shallow-water, near-shore gravels comprising rounded river and beach pebbles and cobbles (conglomerate), followed systematically by sands (sandstone), silts (siltstone), muds (mudstone),

4. Charles Fleming's famous 'paleogeographic' maps were the first attempts to depict the changing shape of the New Zealand land surface through the past 56 million years of Cenozoic time. When first published in the 1960s, they were inspirational, and they were soundly based on the known geology of New Zealand.

marl (carbonate-rich mudstone), glauconite-rich sediments (greensand) and biogenic carbonate sediments (limestone).

This sequence is typical of the effects of an environment with a diminishing sediment source area, an increasing distance from land and an increasing water depth. The transgressive sequence invariably culminates in sediments of biogenic origin (that is, produced by living organisms) and/or authigenic origin (formed 'in place'), which are almost devoid of any clastic input (that is, sediment derived from pre-existing rock). The absence of clastic sediment suggests an absence of land.

What this means is that the widespread limestones of the middle Cenozoic (Late Oligocene to Early Miocene time, 29–16 million years ago) represent a time of maximum flooding of Zealandia by the sea. On the basis of this evidence, we can be sure that this was a time of minimal land area in Zealandia. This has long been recognised by New Zealand geologists, but just how much land was there at the point of maximum inundation?

Depicting How Much Land Was There

The first efforts to tackle the question of the land area of Zealandia were made by Charles Fleming in the late 1950s. Biologists concerned with the origin of New Zealand's plants and animals urged him to address the matter. As an outstanding geologist and paleontologist, with very broad interests and knowledge of New Zealand's natural history, he was perfectly placed to attempt this task. As a result, he produced the first attempts to depict the extent and shape of the New Zealand land area through the Cenozoic Era (65–0 million years). He presented a series of six cartoon maps showing how the shoreline has changed from Eocene (56 million years ago) to Pleistocene time (1 million years ago).

This proved to be quite momentous. It showed a greatly diminished land area during Oligocene time (33.7–23.8 million years ago), and biologists were quick to realise the implications: it helped explain why the New Zealand native fauna and flora was so peculiar compared with elsewhere in the world. Not only was New Zealand a remote ocean archipelago, but it had been greatly reduced in size during Oligocene time. Surely this might explain some of the mysterious absences and depletions in the modern New Zealand biota?

Exploring Flatness

Over 40 years have elapsed. Fleming's contribution and insight has lasted well and his maps have been, and still are, widely used and supported. However, in the mid-1990s, Otago geologist Chuck Landis made some startling revelations. He and colleagues began to question the long-held 'peneplain' explanation for the origin of extensive flat or near-flat surface features that are so well preserved in the bare, tree-less, expansive topography of central Otago.

A 'peneplain' is considered to be a surface produced by the long-term wearing down of the land by terrestrial processes. There was an ancient land surface in Otago in Late Cretaceous time, when Zealandia departed Gondwanaland. But what Chuck Landis was able to demonstrate is that the so-called 'peneplain' is in fact much more complicated. It is a Cretaceous land surface that has been modified by marine processes, cut by the sea, at least twice, since Late Cretaceous time.

Although hard to comprehend, it makes perfect sense when you subtract the regional effects of subsequent tectonism and erosion of the past 23 million years (Miocene to Recent). In places, there are tell-tale remnants of the transgressive marine Zealandian cover sediments that support this new explanation. In the main, however, within central Otago the Zealandian cover sequence has been stripped off and all that is left is the marine-cut surface developed on hard basement rock (greywacke and schist).

Now that we are much more certain about the marine origin of these flat surfaces, they can be tracked throughout the New Zealand landscape and reinterpreted in terms of the submergence history of Zealandia.

5. Chuck Landis (University of Otago), sedimentary geologist and a leading intellect within New Zealand geology. He has one of the first to embrace plate tectonic theory in the 1970s and in more recent years has challenged conventional understanding of the origin and antiquity of the New Zealand land surface.

6. This remarkably flat surface is the remnant of a much more widespread surface that was cut by the sea in Eocene time as Zealandia sank, and was subsequently buried by marine sediments including limestone of Oligocene age. New Zealand has since risen and, as luck would have it, this particular area on Mt Misery, in north-west Nelson, has escaped erosion.

Questioning the Basis for Fleming's Shorelines

More recent research by Chuck Landis and his colleagues has done just this. They have now explored the whole of New Zealand, looking for hard evidence of land during latest Oligocene to earliest Miocene time. In so doing, they have carefully examined the geological basis for the location of the shoreline as depicted by Fleming. The results are extremely interesting and have profound implications for our understanding of the antiquity and origin of the modern New Zealand landscape and its native biota.

The results of this research show that the geological evidence for terrestrial conditions 23 million years ago, at the time of maximum submergence of Zealandia, is very weak. It is so weak that we can no longer assume that there was a continuous land area across latest Oligocene to earliest Miocene time. There may have been, but we cannot demonstrate that there was on geological evidence.

More importantly, if continuity of land cannot be assumed, then we are logically forced to consider the alternative: the complete submergence of Zealandia. The case for complete submergence cannot be proved, but it is just as strong as the case for the presence of continuous land.

In the absence of any substantive geological constraints for the placing of a shoreline anywhere within Zealandia 23 million years ago, the inevitable conclusion is that it was empty, open ocean. What a possibility!

Where does this place us? What does it mean? It means that we have a new theoretical basis for a fresh examination of the geological evidence and mechanisms governing the origin of New Zealand's native plants and animals. These ideas are testable. They demand much more rigorous investigations of latest Oligocene to earliest Miocene rocks of Zealandia, and especially any terrestrial evidence.

The Sinking Mechanism

So much for the rock record and its limitations, or otherwise. What about a mechanism for the sinking of Zealandia? Does this scenario make sense in terms of crustal processes?

Well, our geophysics colleagues wonder what all the fuss is about! They have no problem with Zealandia sinking. To them the evidence is clear enough: the great bulk of Zealandia is submerged and at an ambient level of repose some 1000–2000 metres below sea level. The rifting process commenced 83 million years ago and Zealandia was moved away from the thermal rift that caused Zealandia to split from Gondwanaland. As it moved north-eastwards, it cooled, resulting in loss of buoyancy. It was also stretched and thinned, and as it did so it lost further buoyancy. So, no wonder today Zealandia is largely submerged. The rifting resulted in the formation of the Tasman Sea floor and lasted only about 20 million years. Since then, about 63 million years ago, Zealandia has been more or less locked with respect to the Tasman Sea floor and Australia.

Much of this explanation relates to processes operating within the mantle. Modern research has now established that mantle convection can generate vertical changes in elevation of the Earth's surface with a wave-length amplitude of about 2 kilometres. In other words, the crust can rise and fall by 2 kilometres as a direct result of flow within the mantle. This figure is entirely compatible with the 'height' of Zealandia. So, conceptually, the suggested geophysical mechanism for the sinking of Zealandia is entirely plausible.

Zealandian Islands?

If Zealandia was completely inundated 23 million years ago, could there have been islands? The answer is that there may have been some islands. However, there is no geological evidence for any. We can neither confirm nor deny on the basis of geology. If there were, they must have been short-lived, and left no geological trace.

Biological arguments can be made for the continuous presence of islands. There is a suggestion that the ratite ancestors of the moa in particular, but also the araucarian ancestor of the kauri, and some other primitive organisms such as the tuatara and leiopelmatid frogs, must have existed somewhere on Zealandia in splendid and sustained isolation from Gondwanaland land masses such as Australia and Antarctica. Similarly, a case can be made for islands, because many of the plants that subsequently colonised New Zealand appear to have been pre-adapted to an island lifestyle, as opposed to a continental lifestyle.

If we look around greater Zealandia today, most of the major islands, such as Stewart Island and the Chatham Islands, were completely submerged

23 million years ago, or simply did not exist, especially volcanic islands such as the Norfolk, Lord Howe, Auckland and Campbell Islands. New Caledonia, located on a northern promontory of Zealandia, became emergent in Oligocene time, less than 10 million years before New Zealand did.

The Timing of Maximum Immersion
How do we know the age of maximum immersion of Zealandia? From marine fossils preserved in the limestone, and in particular, from microfossils, especially fossil plankton: foraminifera and nannofossils. There are also distinctive, larger shelly fossils including bryozoans, corals, brachiopods, molluscs, barnacles, echinoderms and vertebrates (whales, penguins, fish, turtles), that enable us to be certain of the age of deposition.

Limestone is biogenic sediment – that is, of biological origin. In other words, the sediment that makes up limestone is entirely comprised of the skeletal remains of shelly organisms. It is wall-to-wall fossil! The Late Oligocene to Early Miocene limestone of Zealandia is no exception. It was deposited in an environment that was virtually rock-free. The fine matrix of limestone is in part carbonate mud and silt derived from the mechanical erosion of shelly debris, and in part from natural carbonate cement that forms and sets during the burial history of the limestone.

Limestone can form in freshwater lakes, but the Late Oligocene to Early Miocene limestones of Zealandia are all marine. You can tell this at a glance just by looking at the fossils. The diagnostic remains of sea creatures are a giveaway.

7. The 'pancake rocks' at Punakaiki (West Coast, South Island) are a major tourist spot and they owe their attraction to the seaside location. The sea has picked the limestone clean to reveal their remarkable structure, and they do indeed resemble a great stack of pancakes. The key to understanding this is pressure solution. The limestone, of Oligocene age, has been loaded and squeezed resulting in dissolution along a set of regularly spaced surfaces within the limestone. These surfaces are the boundaries between the 'pancakes'.

8. Limestones of Oligocene to Miocene age in New Zealand are characterised by an abundance of fossil debris derived from echinoderms. Complete specimens of heart urchin, such as this exquisite example of *Pericosmos crawfordi* are not uncommon, and are eagerly sought after by fossil collectors.

What Did the Limestone Environment Look Like?

It was open sea, possibly devoid of islands. Imagine a shallow sea with water depths of no more than about 200 metres. Zealandia would have presented a vast submarine platform teeming with life.

Over a period of about 10 million years, limestone accumulated and formed a blanket of sediment tens of metres thick, but probably less than 100 metres thick. If you think of limestone as white, then Zealandia would have resembled a large, irregular-shaped cake with icing on it! It would have been underwater, of course.

It is sobering to think that this 'rock' was entirely produced from the skeletal remains of living organisms. Knowing the area of Zealandia, and the thickness and density of the limestone, you can work out the weight of skeletal material involved and figure out just how productive the Pacific Ocean was above submerged Zealandia in terms of biomass. It is mind-boggling!

From all that we know of the fossil composition of the limestone, we can be certain that the environment was not tropical. It approached subtropical conditions but was largely of a warm temperate. There were no coral reefs, for instance. There were solitary corals, but no reefs. The waters were warm enough for abundant life, dominated by sea-floor forests of bryozoans and abundant sea urchins (echinoderms), but were too deep for coral reef formation. And it was not so deep as to be unaffected by strong currents and waves that swept the sea floor. Most of the limestone accumulated between water depths of 50 to 200 metres.

Pancake Rocks at Punakaiki

One of the better-known limestone localities in New Zealand are the Pancake Rocks at Punakaiki on the West Coast of the South Island. These rocks are unusual because they have been subject to pressure solution. They have been loaded with such a pile of sediment that the limestone has started to dissolve along a set of closely spaced, sub-horizontal surfaces within the rock. This 'dissolution' can only happen in the presence of fluids, but it happened!

Pressure solution is a completely natural process best illustrated by using an analogy of an ice skater. The weight of the skater exerts intense pressure on a very small linear contact area on the ice. The pressure is such that the ice is instantaneously melted immediately beneath the blade of the skate. There is a direct relationship between temperature and pressure. Intense pressure raises the temperature, hence the localised melting of the ice, and the production of water beneath the blade lubricates it and greatly assists the speed of the skater.

A close examination of the Pancake Rocks, especially between the 'pancakes', shows a hairline suture pattern, not unlike the highly convoluted suture lines on a human skull. This represents just one surface of dissolution. Within the pile of limestone at Pancake Rocks, there are thousands of dissolution surfaces, and each one represents loss of rock. The dissolution or dissolving of carbonate took place at a depth of hundreds of metres below the sea floor. The limestone is muddy and the dissolution process has effectively removed carbonate and concentrated the mud component. This is why the pancakes have a muddy look about them, and it is probably the distribution of mud that controls the thickness of the pancakes.

The pancake effect is entirely secondary, but, as luck would have it, the pancakes are sub-parallel to the original bedding or sedimentary layering of the limestone. The pancake structure therefore looks as if it is primary. It is in fact just a happy coincidence. Incidentally, the dissolution process is no longer happening: it occurred only as a function of burial. The cover rocks that buried the limestone and caused the dissolution process have now been eroded off because of subsequent uplift due to plate collision.

The Pancake Rocks are an unusual example of Zealandian limestone that

9. Limestone country in Castle Hill Basin, inland Canterbury, is another erosional remnant of the substantial blanket of limestone that draped much of submarine Zealandia during Late Oligocene to Early Miocene time, testimony to widespread if not wholesale inundation by the sea 23 million years ago.

was loaded by New Zealand sediment. They are New Zealand pancakes made from Zealandian ingredients! But for the cooking effects of plate collision, they would not be. The strip of land that we know of as the West Coast of the South Island owes its existence in large part to the proximity of the Alpine Fault to the plate boundary. The Pancake Rocks are just one example of many 'local' effects.

Another local effect is the special quality of Zealandian coal. This is greatly prized for specific industries and coal mining is the biggest economic activity on the West Coast. Yet the coal owes its special attributes once again to being 'loaded and cooked' as the continental plates collided. In places it has been caught up in the collision zone itself and folded and/or rotated from its original horizontal aspect into the vertical. Some coalmines are exploiting vertical seams of coal.

For the most part, the Oligocene to Early Miocene limestone occurrences are remarkably undeformed and are almost flat-lying, just as they were when they accumulated. Erosional effects have produced some spectacular limestone landscapes in many parts of New Zealand, from Southland to Northland, and none more so than in north Otago and south Canterbury. Although it would have been buried by younger Miocene to Pleistocene sediments – all the material deposited since 23 million years ago – the limestone was never buried by more than a few hundred metres.

Caves

Some exposed limestone surfaces are studded with classic sinkholes. These strange features are due largely to chemical weathering, or natural dissolution caused by slightly acid rain, soil and groundwater effects. Some surface features are also caused when limestone caverns collapse deep underground. What can be seen at the surface often belies a much more eroded subsurface.

New Zealand limestones are well known for their caves, including those

10A

10B

at Lake Te Anau, the Nelson area and the Waitomo Caves near Hamilton. All these cave systems have come about as emerging New Zealand has superimposed itself on older Zealandian limestone.

What is interesting is that a number of very important fossil deposits, primarily of bird bones, have been found within alluvium in caves. Yet no cave deposits have been discovered as yet in New Zealand that are greater than 1.8 million years old: they are all of Pleistocene age, and usually much less than 1 million years old. It would be very exciting indeed to find older Pliocene (1.8–5 million years old) deposits, but perhaps they simply do not exist. This could again be indicative of the remarkably youthful uplift history of New Zealand.

Castles of Limestone

There are a number of places in New Zealand where there are remnants of the limestone forming upstanding cliffed topography, reminiscent of the ramparts of a castle. Hence the names: Castle Hill Basin on the way to Arthur's Pass in Canterbury, and Castle Hill in the Taringatura Hills of Southland. These, too, are sculptured from limestone, but in both cases the limestone has been warped or folded by tectonic effects as plates collided. Note that Castle Hill near Wanganui and Castlepoint on the Wairarapa coast of the eastern North Island are different; their names do not relate to the limestone.

One of the great joys of the limestone is searching for beautiful fossils. Some of the best-preserved and largest fossils to be found are heart urchins, and they are often intact. These echinoderms are relatives of our modern sand dollars. Solitary corals and beautiful ornate gastropods are also easy to find, as are shark teeth.

Oamaru Stone

For simplicity, we have referred to the limestone as if it were one entity, but from a geological mapping perspective there are many limestone formations. They all have different names, and each relates to a particular area. Understandably, they vary in time and space. Oligocene time lasted about 10 million years (34–24 million

10. A & B: Not just a building stone, Oamaru Stone is a limestone of Late Oligocene to Early Miocene age. It accumulated as biogenic sediment in shallow sea during maximum submergence and flooding of Zealandia. It is the remnant of a much more widespread limestone formation. The principal ingredient of this limestone is derived from bryozoans. These are colonial animals that produce calcite skeletons in many different shapes and sizes, as illustrated here.

years ago), but limestone deposition in Zealandia began in Late Eocene time and continued into Early Miocene time (37–16 million years ago). Inevitably, the limestone varies enormously in character.

Perhaps the most famous Oligocene limestone is Oamaru Stone, referred to as Ototara Limestone by geologists, and it is of Early Oligocene age. This is quarried on a relatively large scale near Oamaru for both agriculture (fertiliser) and the construction industry. It makes a handsome building stone and is also much used for decorative sculpting. What is interesting is that it is comprised largely of the skeletal remains of bryozoans. Other shelly fossils, such as brachiopods, scallops, fossil teeth and bones, are conspicuously present, but collectively these fossils are almost insignificant compared with the extraordinary abundance of bryozoan remains.

Whale Fossils

Areas of Late Oligocene to Early Miocene limestone occur in many places throughout New Zealand, and most areas support major quarries for the cement and agricultural lime industries, but the by far the best places to see it are south Canterbury and north Otago. This region of the South Island has proved to be a bonanza for whaling paleontologists: those who study fossil whales. Skeletal remains of such large animals are not uncommon, but it is rare to find (let alone collect) a whole skeleton. Normally, a scatter of bones is found, as if strewn over the ancient sea floor and buried in sediment. The most diagnostic part of a skeleton, any skeleton, is the skull, and within the skull, of a whale at least, are the ear bones. These are the bits of greatest diagnostic value in terms of trying to determine what sort of whale the fossil relates to.

There are two major groupings of whales: baleen whales and toothed whales, including dolphins. Fossils of both groups have been collected. Some complete skulls of toothed whales have been found, including superb specimens of *Waipatia* and *Squalodon*, described by Ewan Fordyce.

Ewan Fordyce

For decades now, Ewan Fordyce (University of Otago) has systematically explored the limestone for whale fossils. In his quest, he has discovered many other fossils of interest, not least the remains of penguins, turtles and large fish, including sharks. These immensely exciting finds often require a huge amount of technical effort to extract and prepare them for examination. The thrill of the chase is sustained as the fossils are slowly excavated from their rock matrix, often to reveal something completely new to science. The Geology Department at Otago University is now brimming with large fossils collected from the limestone. Then, as word gets out, there is an instant demand for replicas of the more showy and spectacular finds.

All this concerted effort has resulted in a remarkably thorough knowledge of the larger fossils preserved in the limestone. It has been coupled with a great deal of research on other aspects, such as the history of sea-water chemistry during Late Oligocene to Early Miocene time. Remember that the limestone is comprised of calcium carbonate skeletons grown by organisms living in sea water, so their skeletons are a chemical record of the prevailing environmental conditions, especially the strontium isotopic composition of sea water, and the isotopic signature of oxygen and carbon. Much can be deduced from these chemical and/or isotopic signatures, especially water temperature. The most useful fossils are planktic foriminifera.

Hunting for fossil whales in north Otago and south Canterbury has produced a significant by-catch of fossil penguins, but Oligocene penguins are not restricted to the South Island. In 2006 a naturalist club based in Hamilton made the news when they found a spectacular, near-complete fossil skeleton of a large penguin in Kawhia Harbour.

Few other bird bones are known from Oligocene rocks, but a single fossil wishbone is thought to be an albatross. Bird bones are lightweight and generally fragile, so it is hardly surprising that so few are known. Only the most robust bones survive the destruction on the sea floor. Penguin bones are much more robust than those of most of the ocean-going marine birds such as petrels, shearwater and albatrosses.

Conclusion

In this chapter we have considered the geological evidence and thinking for saying that Zealandia sank, and to what extent it sank. The widespread occurrence of limestone of Late Oligocene to Early Miocene age is especially relevant to this interpretation, but in itself does not preclude the continuous existence of land. So, while it cannot be conclusively shown that Zealandia was completely drowned, nor can it be shown that it wasn't. Therefore both scenarios must logically be explored. Geologically, we can no longer assume that there necessarily was continuous land in the New Zealand region of Zealandia.

The next chapter considers the emergence of New Zealand, as we know it today.

11. Not to be messed with. A single tooth of *Carcharodon megalodon*, a gigantic ancestor of the Great White Shark of today. This was found in limestone of Miocene age from Cape Foulwind, near Westport.

12. The skull and jaw of *Squalodon*, another spectacularly toothed whale fossil, this time collected from Oligocene limestone from North Otago, near Duntroon – pictured here with Andrew Grebneff (University of Otago).

13. Ewan Fordyce is New Zealand's leading vertebrate paleontologist with specialist interests in the paleobiology and evolution of whales, and the communication of science to the public. Here he is examining a Cretaceous mosasaur fossil from Shag Point, near Palmerston.

Part 04:

New Zealand: 23–0 MILLION YEARS AGO

13/ The Emergence of New Zealand

The quiet stability of submerged Zealandia was suddenly arrested from its sleepy repose. Almost as if a switch had been turned on, the boundary between the Pacific and Australian plates became vigorous, cutting a fresh pathway right through Zealandia. The stirrings of this development can be traced back to about 45 million years ago during Eocene time, but it really took off 26–22 million years ago. Prior to

The Emergence of New Zealand

1

1. A winter scene of New Zealand from space with White Island erupting. New Zealand may be thought of as the tip of a continental iceberg, the emergent part of Zealandia, a sunken continent.

45 million years ago, the plate boundary lay far to the east and was relatively passive. It had no kick. However, it found its feet, and rapidly so, in earliest Miocene time.

This dramatic development caused multiple effects throughout the New Zealand region of Zealandia, with the start of subduction-related vulcanism, the development of the Alpine Fault, and mountain building that produced the axial ranges of the North Island and, of course, the Southern Alps. These are the most conspicuous effects, but most importantly it generated a great upstanding welt in the Zealandian crust: New Zealand. It was a land oriented north-east to south-west, sub-parallel to the plate boundary collision front, and at least 1000 kilometres long and 100 kilometres wide.

The collision was manifest, not unlike the present situation, and it has been sustained ever since. The result is New Zealand as we now know it, risen up from sea. But how do we know that this is what happened? How certain can we be? This chapter explores some of the evidence.

A Flood of Sediment

As if the tape of geological history was being rewound, the Early Miocene to present (23–0 million years ago) rock record is a reversal of the Late Cretaceous to Early Miocene transgressive marine sequence. This time, the sequence starts with limestone and greensands associated with maximum inundation of Zealandia 23 million years ago, then passes upwards into marls, muds, silts, sands and gravels. Understandably, this sequence is referred to as a 'marine regression sequence'. It is a record of shallowing of the sea, a decrease in distance to land and a substantial increase in land area, this time with pronounced topography.

The change was dramatic, as is well demonstrated by the huge and extensive volume of Miocene clastic sediments (that is, taken or broken from pre-existing rock) that dominate the on-land marine basins of New Zealand, and, more significantly, the submarine basins of submerged Zealandia. The volume of sediment records a sudden rise of the land, followed by steady, sustained uplift and consequent erosion.

From Passive Margin to Active Margin

Geologists attribute this change to the vagaries of plate tectonics, which, as stated before, is entirely controlled by convection processes in the mantle. In terms of the plate boundary, the switch is referred to as a change from a 'passive' margin to an 'active' margin. The two differ enormously. To be specific, the action is collision (that is, compression), as opposed to extension or rifting, whereby land areas become non-tectonic.

The Oldest Terrestrial Fossils of New Zealand

In recent years some of the oldest known terrestrial fossils from New Zealand have been found in lake sediments from central Otago. These sediments are considered to be 19–16 million years old, so they are of Early–Middle Miocene age, and they postdate the maximum flooding of Zealandia by some 4–8 million years (the exact age of these sediments remains imprecise, so we must express the age as an age range). What this means is that, from all available data, the age of these sediments is unlikely to be younger than 16 million years and unlikely to be older than 19 million years.

They are affectionately known as the 'Lake Manuherikia' sediments by geologists, and remnants of these rocks can be found over a very wide area of central Otago. The ancient Lake Manuherikia must have been about twice the size of Lake Taupo. The sediments include gravels, sands, silts, muds and coals. The coals are of low grade because they have not been buried very deeply, but nevertheless there are coal mines near Roxburgh and also the Ranfurly region of the Maniototo Plains that have long been operating on a small scale.

These Manuherikia rocks must have accumulated in a large, shallow lake complex, fed by streams draining uplands to the west. They thus represent a relatively early chapter in the history of New Zealand, as opposed to Zealandia. These rocks, and the fossils they contain, are especially deserving of more intensive research, as they offer the promise of much fresh knowledge about life on newly emergent New Zealand.

St Bathans Bonanza

It is hard to imagine New Zealand with freshwater crocodiles, but they were once here. Fossil crocodile bones and teeth have been recovered from several 'bone beds' in the Lake Manuherikia sediments near St Bathans. These are layers of lake sediment with a high concentration of bone fossils, mainly of fish and waterfowl. They are of huge interest, so substantial collections have now been made. However, it is painstaking and laborious work to extract individual bones and make sense of them. Many are small and broken, and are not well preserved. But some wonderful finds have been made that indicate a much more diverse fossil record than fish and ducks.

The fossils include bones and fragments of egg shell of at least 25 species of bird, including probable moa, several species of lizard, tuatara-like sphenodontid fossils, four species of bat, freshwater molluscs and land snails. The crocodile bones are particularly interesting, as are a number of other extraordinary discoveries, including ground-dwelling mammal fossils.

2. Zealandia was largely submerged 23 million years ago. Then New Zealand started to rise out of the sea. This map shows the distribution of Miocene sediments in New Zealand, most of which are marine.

3. Although foreign to most New Zealanders, this is the unmistakeable fossil bone of a small freshwater crocodile, collected from lake sediments near St Bathans. It is most likely 19–16 million years old. Crocodile teeth and bones were first recognised from this site by paleobotanist Mike Pole, but more fossils have subsequently been found.

4. The St Bathans area, central Otago. The widespread occurrence of fossil lake sediments tells us that in Miocene time, 19–16 million years ago, much of this region was the site of a large lake, referred to by geologists as 'Lake Manuherikia'. Fossil evidence, first discovered by Dunedin geologists Barry Douglas and Jon Lindqvist, tells us that it abounded with waterfowl, fish and freshwater crocodiles. Geologists have long been interested in these lake deposits because they are coal-bearing and have been actively mined in places, such as in the Ida Valley and Roxburgh.

This fossil bonanza is now recognised as one of the richest known fossil bird localities globally, and is now the focus of international interest. It rates as one of New Zealand's most significant fossil localities. For this reason, it needs to be protected and conserved in such a way that no potential scientific information is lost.

New Zealand's Earliest Mammals?

Several tiny bones have been attributed to non-flying land mammals. One of these bones is a fragment of a jaw with 10 tooth sockets, and yet it is less than 6 millimetres long. Two other fragments of limb bone are considered also to be mammalian, and may or may not relate to the jawbone. But there is little doubt they are mammal bones. What is even more interesting is that the jawbone fossil is attributed to a primitive group of mouse-sized animals that can be traced back to the Early Cretaceous (up to 146 million years ago), predating the separation or divergence of marsupial and placental mammal groups. It seems that the St Bathans mammal harks back to an ancient Gondwanan lineage.

These discoveries are momentous: they appear to represent the earliest known, truly land-dwelling New Zealand mammals. Until then, the oldest, truly terrestrial mammal fossils known from New Zealand were Pacific rat bones (kiore), but these are considered to be only about 700 years old. It is now clear that kiore arrived in New Zealand with Maori. Note that rare, older bat fossils (more than 700 years old) have been recorded, and there is a reasonably good record of marine mammals (seals and whales) for much of New Zealand's 23-million-year history.

To hunt for fossils

A great deal of our knowledge about the ancient history of New Zealand and its life depends on fossils. It is most satisfying to find a spectacular fossil, or, even better, a spectacular fossil that is also of great scientific significance. However, it isn't all that easy to do so.

You have to know your rocks well enough to know where not to look. Fossils generally occur only in sedimentary rocks, and are rarely found in igneous and metamorphic rocks. You also have to have an eye for fossils, and it helps if you know what you are looking for, and what to expect. Of course, you get better with experience. It's a bit like shopping for bargains.

There is a widely held misconception that paleontologists dig for fossils, but digging is the exception rather than the norm. Most fossils are found by keen observation of rock surfaces. What draws the eye is a subtle change in colour, texture and/or shape, against the ambient sediment background. Once spotted, you are captivated. Normally only a small part of the fossil is revealed on the natural surface of the rock, so you then need to carry out some excavation work in order to extract the fossil. Depending on the strength or hardness of the rock or sediment matrix, geologists and paleontologists use a geological hammer as a matter of choice, often with the aid of a cold chisel. These items are stock-in-trade for fossil hunting. If the sediment is soft, a conventional garden spade is the weapon of choice, or a flat-bladed knife and a stiff hand brush.

Every paleontologist has his or her speciality. If you study fossil whales (like Ewan Fordyce at the University of Otago) you have very different collecting requirements from someone who studies fossil pollen, or plankton. Perhaps the most common fossils are shelly fossils, and these tend to be the easiest to collect.

In paleontology, there is a distinction between 'macrofossils' (fossils that you can see and hold), and 'microfossils' (fossils that require a microscope in order to see them). Within the category of 'microfossils', there is a distinction between 'organic walled' microfossils, and 'mineral walled' microfossils that are usually calcareous or siliceous.

Some people are better at spotting fossils than others. If you happen to find a fossil, and you are unsure of its significance, it is worthwhile contacting someone who might be able to help. In this way, members of the public have greatly assisted science in New Zealand. Many fine fossils have been found by farmers and fishermen over the years, and many have drawn their discoveries to the attention of geologists, for instance: the large collection bequeathed to the University of Otago by John Graham.

i. To search for fossils is to search for hidden shapes, but not usually by digging. Fossil hunters train their eyes to spot unusual patterns, textures, shapes and colours on rock surfaces. This is a boulder in Mangahouanga Stream, inland Hawke's Bay, with a conspicuous object in the rock: fossil bone. The next step is to collect the boulder or a piece of it, and try to extract the bone. Who knows what else is inside this rock! Only careful preparation will reveal all.

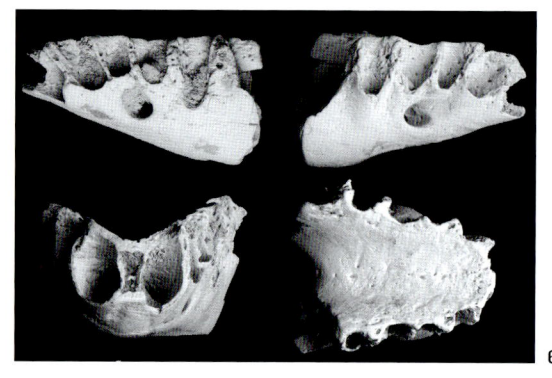

5. The tuatara (*Sphenodon punctatus*) is one of New Zealand's strangest and most mysterious residents. In biological terms it is half-dinosaur, half-lizard, and belongs to a group of reptiles that can trace its origins back to Triassic time. However, this does not necessarily mean that the tuatara has been in New Zealand since Triassic time! The fossil record of tuatara is very poor indeed, with few, if any, pre-Pleistocene records (older than 1.8 million years). If Zealandia was completely submerged 23 million years ago, then like all other terrestrial life, it has got here since then from somewhere else . . . somehow.

6. New Zealand's first pre-Pleistocene mammal fossil (older than 1.8 million years). This was discovered by professional vertebrate paleontologists and described by the same team: Trevor Worthy and colleagues. They described three small bones, each less than 6 mm long. Two are broken limb bones (femurs) and the other is this unmistakeable jawbone with conspicuous mammalian dental sockets. No teeth have been found . . . yet! It is thought that the animal is of a very primitive mammal lineage, and was mouse-sized.

7. New Zealand has only two native land dwelling mammals and they are both bats: the long-tailed (*Chalinolobus tuberculatus*) and smaller short-tailed (*Mystacina tuberculata*) shown here.

8. Exquisite fossil preservation of a scale insect within a leaf. Yet it is ancient, of Miocene age, somewhere between 20 and 15 million years old. This was found by Daphne Lee and Jennifer Bannister in diatomite at the Middlemarch Maar. For preservation of such delicate material, exceptional natural conditions are required: no oxygen, no bacteria, no natural radiation (light) and no mechanical agitation. This leaf, and its insect occupants, must have fallen into relatively deep, dark, cold and still lake water, almost totally devoid of oxygen and bacteria. The preservation is quite extraordinary.

The fact there were Miocene mammals in New Zealand will mean a rethink of the origin of the present New Zealand native fauna and flora. In the absence of any fossil evidence of mammals, it has long (and understandably) been assumed that our wonderful iconic New Zealand birds and plants have evolved pretty much free of mammals, except for bats. All sorts of interesting questions flood to mind and will occupy the thoughts of natural historians for years to come.

For instance, if we can be certain that there were mammals here 19–16 million years ago, what happened to them? Where did they come from? When did they arrive? Were they always here? Does their existence mean that Zealandia was never fully submerged? How diverse were these mammals? Were there others? Was there ever a mammal megafauna, in other words large mammals comparable in size to today's mammals? Were there marsupials?

One of the obvious observations from all this newly acquired knowledge and understanding pertains to the fossil record itself: how fickle it is! We have been in the dark for so long, and all because of a want of key fossils. The fossil record can be very misleading because it is demonstrably so incomplete. As already stated: one bad fossil is worth more than a good theory. Fossil hunting is therefore a very worthwhile exercise.

Discovering the St Bathans Bonanza

Two Dunedin-based geologists, Barry Douglas and Jon Lindqvist, discovered the locality near St Bathans in the early 1980s. Paleobotanist Mike Pole found the first crocodile fossils, and described the find with Ralph Molnar in 1997. Subsequently, vertebrate paleontologists have visited the site. A team of New Zealand and Australian researchers, led by Trevor Worthy, have since reported on their initial exciting finds, but much remains to be described and identified formally.

Many people with specific expertise in one particular group of fossils are now involved, including Alan Tennyson, Michael Archer, Anne Musser, Suzanne Hand, Craig Jones, James McNamara and Robin Beck.

The Middlemarch Maar

Otago has other treasures, including fossiliferous lake sediments preserved in a maar near Middlemarch. A 'maar' is a particular kind of small volcanic crater, usually caused by an eruption of basalt, with high-temperature lava, between 1100°C and 1250°C, that explodes violently on contact with near surface groundwater, producing a flaring, trumpet-shaped depression and a raised rim. (The term 'maar' is German, and relates to distinctive basalt craters in the Eiffel region of northern Germany.) In time, the vulcanism dies away and the crater fills with fresh water to form a lake that is more or less isolated from any stream or river. It is a stand-alone, static body of water, an almost closed system that is replenished by rain and/or groundwater. Some of the Auckland volcanoes may be described as maars.

The Middlemarch Maar is of particular interest because it is infilled with a sequence of finely laminated, Early to Middle Miocene lake sediments. The maar is almost 1 kilometre in diameter, and the sediment infill is in excess of 100 metres thick and may be as much as 200 metres thick. The sediment itself is amazing because it is nearly 100 percent biogenic, in this case almost totally comprised of the skeletal remains of freshwater organisms, and in particular diatoms, which are single-celled algae that produce intricate siliceous tests. Only the top 10 metres of this maar lake deposit have been sampled and so far only one species of diatom has been recognised: *Encyonema jordanii*. The fossil remains of freshwater sponges are also preserved, but by far the most significant fossils are of plants and fish. There are numerous examples of extraordinary preservation, with superbly preserved fossil fish, bark, stems, leaves, flowers, fruits, seeds, pollen, spores, fungi and insects. This site is yet another bonanza and is also deserving of a great deal of careful exploration and research. Much can be determined from this locality about life in New Zealand in Early Miocene time.

The lake was surrounded by a forest that was dominated by trees related to the laurel family and would have been a natural trap for falling or wind-blown forest litter. The lake waters must have been sufficiently still and stagnant for anoxic (that is, lacking oxygen) conditions to exist on the lake floor. The lack of oxygen would have inhibited normal decay processes and with time the lake floor would be buried by yet another season or bloom of diatom skeletons. Hence, the exquisite preservation of so many fossils.

The age of the Middlemarch Maar is as yet uncertain, but an age of 23 million years has been obtained from volcanic rock closely associated with the maar. This means that the lake sediments are unlikely to be older than 23 million years old, and could be considerably younger. It is attributed an Early to Middle Miocene age range, 24–11 million years old, but once a crater is formed, it would quickly become water-filled and occupied by a freshwater ecosystem. It must postdate the time of maximum immersion of Zealandia, but not necessarily by very much. The Middlemarch Maar is a superb record of Early Miocene plant life in particular, and complements the St Bathans record of Early Miocene animal life in New Zealand.

Imagine what more may yet be found at this site. The Middlemarch Maar is comparable to the World Heritage Site at Messel in southern Germany, where lake sediments of Eocene age, about 48 million years old, have produced the most amazing fossils with extraordinary preservation.

Scientists at the University of Otago are actively researching the Middlemarch Maar, led by geologists Daphne Lee, Jennifer Bannister and Jon Lindqvist.

Meanwhile, to the North: Allochthons Slide into Place

It is hard to believe, but much of the landscape in Northland and the Raukumara Peninsula of the east coast of the North Island is comprised of foreign rocks. That is, they are rocks that have been emplaced *en masse* from elsewhere. Such rock complexes are known as allochthons from the Greek '*allo*', meaning foreign, and '*chthon*', meaning earth. At the time that this happened, in Earliest Miocene time, and at the time of maximum submergence of Zealandia, Northland and the Raukumara Peninsula were adjacent to each other.

These rocks or allochthons, are the remnants of a huge slice of the ocean floor that was peeled off, and slid, under the influence of gravity, in a southwesterly direction several hundred kilometres from its original location. A series of slices is recognised, that originated from an upstanding submarine rise, and slid down, one on top of the other, to form a pile about 4 kilometres thick. Amazingly, all this action was entirely submarine and it all happened within the space of a few million years, commencing about 23 million years ago. It is calculated that the original volume of rock involved was about 100,000 cubic kilometres.

Since the emplacement of these allochthons, the New Zealand crust has been uplifted and subject to erosion. Needless to say, it took a very long time for geologists to figure out this extraordinary story. The critical information for understanding all this was detailed geological mapping and age determination using fossils. What the fossils were able to show is that each 'slice' or 'nappe', to use an old European term, immediately overlies rocks that are younger than the rocks at the base of each slice. The only sensible explanation for this phenomenon is wholesale detachment and stacking of slices of the sea floor along sub-horizontal faults.

This may well seem far-fetched indeed. However, modern analysis of the sea floor using seismic profiling techniques clearly demonstrates the existence of allochthons up to 10 kilometres in length and about 1.5 kilometres thick, sliced and stacked below the sea floor in water depths of 1500–2000 metres.

What this means is that Northland and the Raukumara Peninsula comprise the remnants of oceanic crust derived from the Pacific Plate that have been obducted onto continental crust on the Australian Plate. (Obduction is the opposite of subduction: it means 'led up onto' as opposed to 'led under'.) It is yet another manifestation of plate collision, and, notably in this case, the onset of vigorous collision about 23 million years ago. Examples of obduction are not uncommon in the geological record, but may nevertheless be considered as a kind of accident. Nature is not perfect!

All of this tectonic action, the emplacement of the Northland allochthon, predates the emergence of land in what is today the North Island.

The Northland Volcanic Arc

Immediately after the emplacement of the allochthon, a volcanic arc became established in Northland. Remnants of a double chain of volcanoes have been mapped, with about 40-kilometre spacing.

9. Daphne Lee (University of Otago), paleontologist with specialist interests in brachiopods, paleobotany and earth science education, and research leader for investigations of the Middlemarch Maar fossil bonanza of Miocene age.

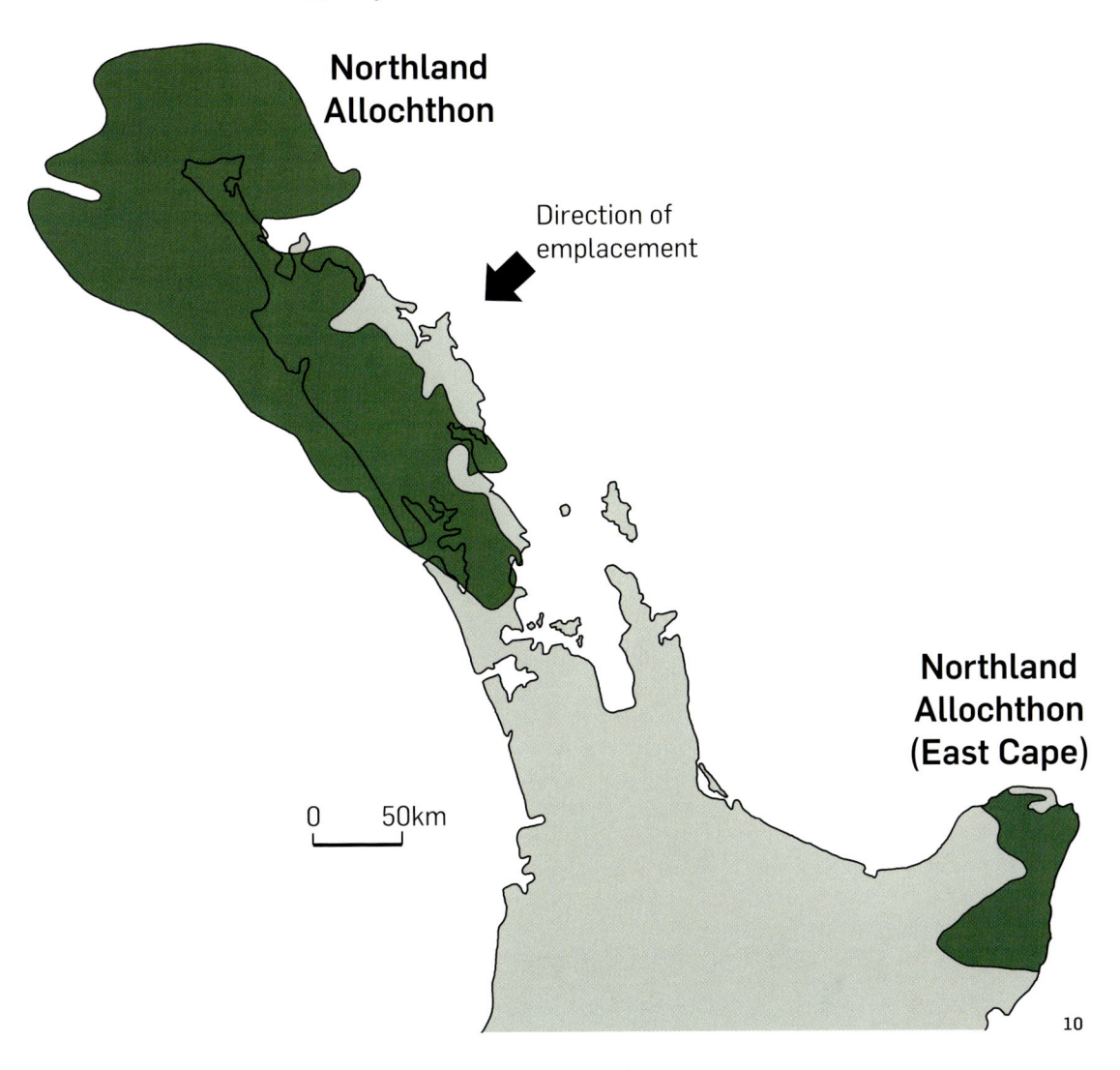

They erupted 20–15 million years ago, caused by the Pacific Plate being forced under the Australian Plate, and were comparable to the modern Tonga–Kermadec arc that runs north-east to south-west on land through the Bay of Plenty as far as Ruapehu. Many of the Northland volcanoes would have approached Ruapehu in size and produced the same type of products: andesite lavas, airfall ash deposits and lahars.

Spectacular rubbly lahars can be observed on the northern coasts of Whangaroa Harbour and Whangaroa Bay. The stumps or remnants of these long-extinct volcanoes tell us that the Pacific Ring of Fire has not always been where it is now. It is a moving entity. It is capable of changing direction or location, as a result of changes in convection within the mantle.

As has been stated before, volcanoes are superb mechanisms for the long-term storage of records or tracts of Earth history. They trap and bury pre-existing rocks by virtue of their voluminous 'products', their sudden outpourings: lava, ash and derived sediments. Thanks to their tendency to construct rapidly, they are instant natural conservators, which is ironic, given their status as hazardous destroyers of the environment.

The Waitemata Group
A southern example of remnants of a large Ruapehu-like volcano is the Waitakere Ranges to the immediate west of Auckland. These rugged hills are the remnants

10. Within the past 25 million years, the New Zealand land mass has acquired a substantial addition to its crustal estate, namely 'the allochthon'. Formed elsewhere, it slid down a submarine slope as a single entity at a time when the North Island was perhaps completely submerged between 24 and 19 million years ago. The allochthon has subsequently been split in two and rotated apart by the growth of the central North Island rift, the Taupo Volcanic Zone. This map shows the present-day distribution of the allochthon, now referred to as the Northland Allochthon part of which is recognised in East Cape.

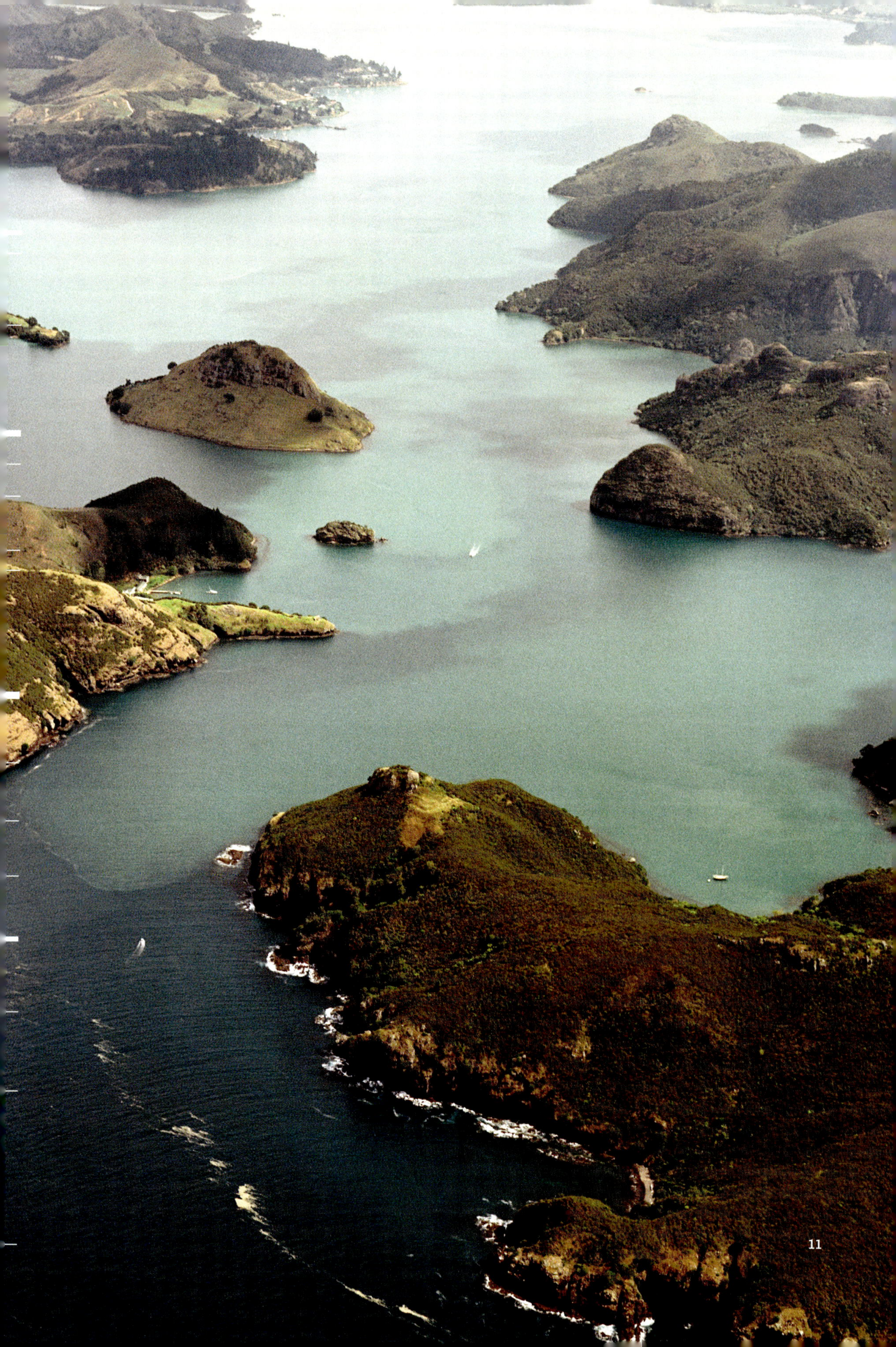

of a large Early Miocene volcano that produced a flood of volcanic debris and sediment known by geologists as the Waitemata Group. These sedimentary rocks can be traced all around the Auckland City region and extend well to the north beyond Kaipara Harbour. They are derived from at least two large, coalescing volcanic complexes referred to as the Manukau and Kaipara volcanoes. What is interesting is that the Waitemata Group is almost entirely submarine. These sediments were eroded off the flanks of largely submarine volcanoes and accumulated in water down to depths of 1500 metres.

Once again, we know this from detailed mapping of the rocks and from careful analysis of fossils. Macrofossils, microfossils and trace fossils enable us to interpret the water depth at which these sediments accumulated.

Waitemata Group sediments are conspicuous in the Auckland City area. They are well exposed on the many beaches and headlands from Tamaki Strait northwards, and can also be seen in road cuts. They have been mapped as far north as Whangarei. The Waitemata Group sediments are the brown-grey-coloured sandstones through which the much younger Auckland volcanoes of Late Pleistocene age erupted.

The Northland arc extended further south and to the west of the present west coast of the North Island. This extension is sometimes referred to as the Mohakatino Arc and is largely hidden from view, eroded and buried by younger sediments and now under the sea. Yet these volcanoes are a significant source of those iconic and conspicuous black-sand beaches of the west coast of the North Island. The black sand is composed of tough, heavy, dense, iron-bearing minerals, mainly iron titanium oxides that have outlasted all less dense and softer minerals. The black sand is all that remains after a lot of natural sieving, sorting and winnowing by the westerly-lashed seas of the Tasman Sea. Note that much younger volcanic rocks of Pleistocene age, less than 1.8 million years old, such as those forming Mt Taranaki, are also contributing to the black sand.

In Early to Middle Miocene time, 24–11 million years ago, the northern North Island must have resembled Vanuatu today. It was a series of largely submarine volcanoes. Undoubtedly, some would have formed islands. They were all located on the continental crust of the Australian Plate, but the trench where the subducting margin of the Pacific Plate plunges down was much closer than it is today, less than 100 kilometres to the east.

However, something else was going on: the entire region was being uplifted because of the plate boundary collision. In the North Island, the collision involved oceanic crust to the west being subducted beneath thin continental crust to the west, but the net effect was one of steady uplift. It commenced about 23 million years ago.

By the beginning of Middle Miocene time, about 16 million years ago, the subduction system was moving eastwards, and the Northland arc volcanoes were slowly switched off. Most of Northland was still submerged, but in water depths of less than 300 metres. Uplift continued, though more slowly, and by the end of Miocene time 5 million years ago, most of Northland was land.

With time, the Pacific Ring of Fire moved eastwards, and during Late Miocene and Pliocene time, 11–5 million years ago, gave rise to the volcanoes of the Coromandel arc, including Great Barrier Island.

Once again, we have been able to figure out this history from detailed mapping of the distribution in time and space, of sediments deposited in the sea, as opposed to those accumulated

11. The remains of an ancient volcano that might have been as large and tall as Ruapehu, now largely eroded away and brought to its knees. The cliffs and hills at the entrance of Whangaroa Harbour (view looking south) are mainly lahars that travelled down the slopes of a large 'Pacific Ring of Fire' volcano about 20 million years ago in Early Miocene time, on a young emerging New Zealand land mass. This volcanic action parallels what is happening today to the east in the Taupo Volcanic Zone, and is caused by the same process: the Pacific Plate diving under the Australian Plate.

12

14

on land. It is especially interesting to note that until about 5 million years ago, there was surprisingly little land in the North Island, and much of it was confined to upstanding volcanoes. The current main axial ranges are surprisingly youthful and have only risen within the past 3 million years. We can be absolutely certain of this because there are remnants of marine sediments, scallop-bearing limestones in particular, on the crest of the Ruahine Range, and there is no doubt about their age: they are less than 3 million years old. This is proof *par excellence* of rapid uplift. As if this were not enough, marine sediments of Pliocene age, 5–1.8 million years old, with conspicuous shelly fossils are readily observed on State Highway 1 between Taihape and Waiouru in the centre of the North Island.

However, parts of Taranaki and the Waikato were land. This we can be sure of because of the presence of lignite and coal of Early Miocene age; coal forms only on land from plant debris from forests and swamps.

And what of Wellington? When did it rise from the sea? The story is much the same as for the rest of the North Island: it is all very youthful. We can be certain that much of it was submerged 5 million years ago.

The North Island is very different from the South Island. Whereas the North is the product of collision between oceanic crust and continental crust, the South is collision between continent and continent. This difference explains everything and, during Miocene time, land was much more extensive in the South Island than it was in the North. Like the Waikato, extensive forest and swamp-lands became established in Southland during Middle to Late Miocene time, giving rise to widespread lignite and coal deposits.

Intraplate Vulcanism On and Beyond New Zealand

It was during Miocene time that major volcanoes erupted to form Auckland Island 19–12 million years ago, and Campbell Island 11–6 million years ago. These islands have existed since only Early to Middle Miocene time, so we can be certain that terrestrial plant and animal life on these islands got there, became established and have subsequently evolved all within the past 20 million years.

The main Dunedin Volcano erupted 16–10 million years ago, although associated sporadic vulcanism in eastern Otago commenced much earlier, 23 million years ago. This was followed by the eruption of the Lyttelton Volcano that formed Bank's Peninsula 12–6 million years ago.

Then the Mangere Volcano in the Chatham Islands fired up about 6 million years ago and produced an island. This would have formed the first land to exist in the Chatham Islands region at the eastern end of the Chatham Rise for some tens of millions of years. We can be sure of this because on Mangere Island there are preserved remnants of a freshwater lake, complete with fossils. However,

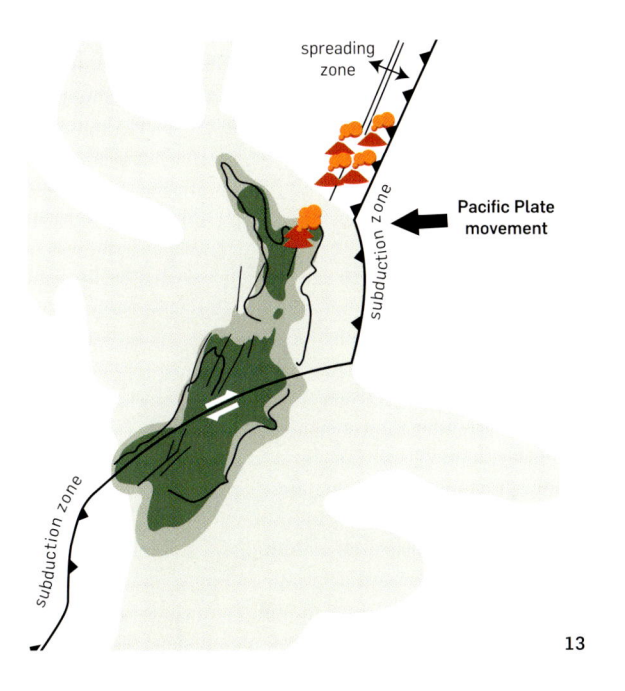

12. Not much left! This is Beehive Island, near Auckland, showing what happens when the sea erodes the land. A flat expansive shore platform has been cut by the waves leaving only a small remnant of the pre-existing topography. The rocks are relatively soft volcanic sediments (mainly sandstone and siltstone) of Early Miocene age. These rocks are widespread in Auckland and Northland. Ironically, they accumulated on the deep submarine flanks of a major subduction related volcano (the Waitakere Volcano) between 22 and 16 million years ago.

13. Five million years ago, at the Miocene–Pliocene time boundary, the North Island looked very different from today. Geological mapping enables us to accurately determine the extent of land.

14. The Dunedin Volcano erupted between 16 and 10 million years ago in Late Miocene time. This view is looking south-west over Otago Peninsula and Otago Harbour. This is the eroded remnant of a substantial volcanic centre, an intra-plate volcano.

15. Eastwards, inland of the youthful active volcano of Mt Taranaki, is a great expanse of marine sedimentary rock, mainly sandstones and siltstones, of Miocene and Pliocene age. Now uplifted and being eroded to produce this confusingly chopped-up landscape, the ridges and valleys stretch as far as the eye can see, all the way to the active volcanoes of Tongariro and Ruapehu in the Taupo Volcanic Zone. The geology tells us that all of this area was under the sea until just a few million years ago.

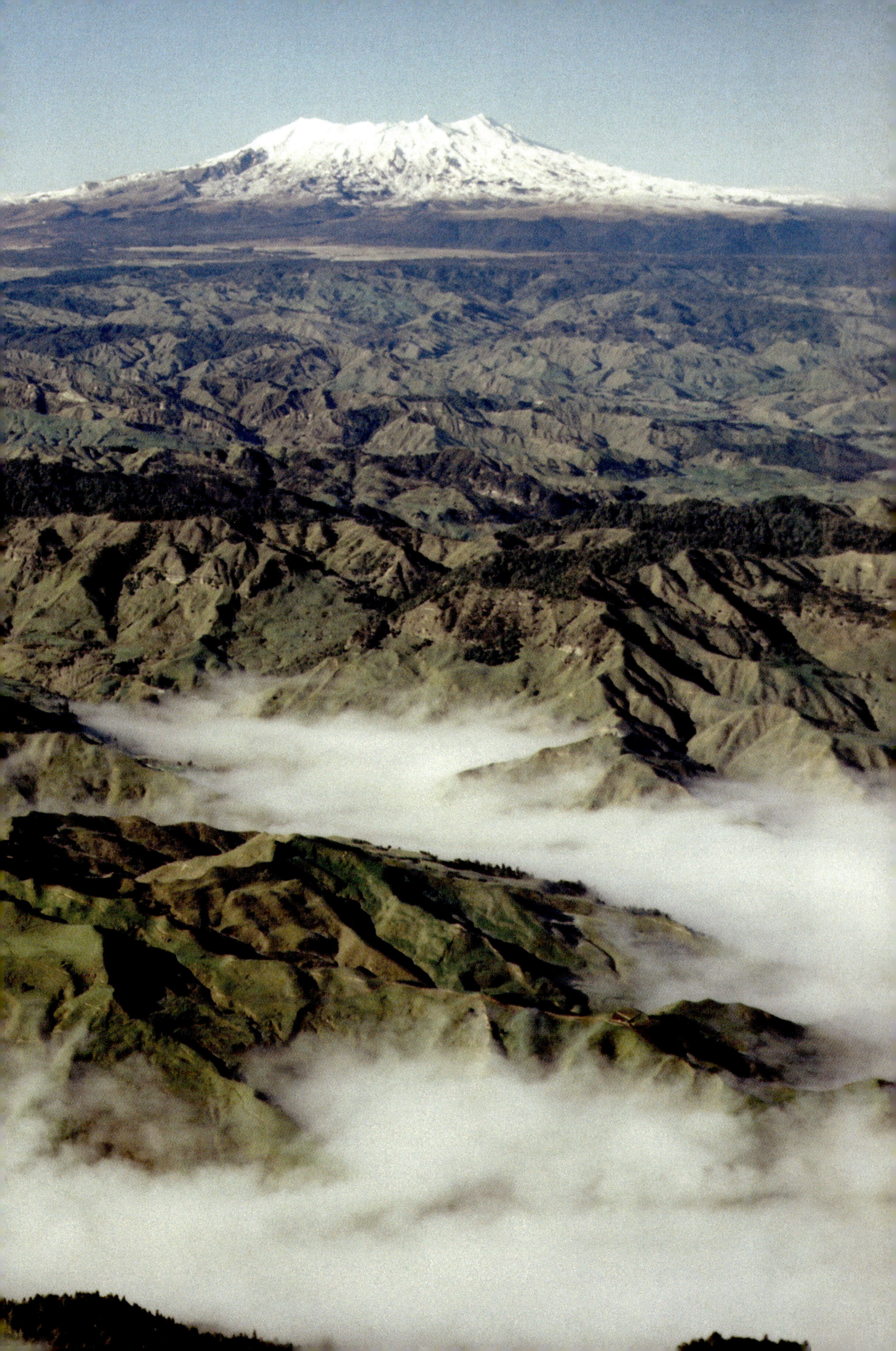

16. Fossil leaves: a sure indication of land, or at least proximity to land. Well preserved leaves such as this one, complete with cuticle, convey considerable information about the environment in which the tree or plant grew. This kind of evidence enables geologists to paint an accurate picture of past life in New Zealand. On the basis of all available Cenozoic plant fossils known from New Zealand rocks of the past 65 million years (including pollen and spores), there is a dramatic change in the composition of terrestrial plant life in earliest Miocene time. The plants that existed prior to 23 million years ago (maximum immersion of Zealandia) are very different from those of today. This specimen is from the Middlemarch Maar locality of Early Miocene age, between 19 and 16 million years old.

this early Mangere island lasted only a few million years. It succumbed to the remorseless sea and was totally submerged and planed flat by wave action by 4 million years ago.

The Miocene Epoch: An Epic Time for New Zealand

The Miocene epoch was the making of New Zealand as a substantial land area, albeit a small part of greater Zealandia. Climatically, New Zealand was generally warmer than it is today and approached subtropical conditions in the north. There were fossil coral reefs preserved in Northland, and we have located fossil crocodiles, coconuts and mangroves in what is now the South Island. Miocene seas were generally about 5°C warmer than they are today. However, this relates to only the warmest intervals of time during the 23–5-million-year time span. Life nevertheless flourished both on land and in the adjacent ocean. If there was no land to speak of at the time of maximum immersion of Zealandia 23 million years ago, then terrestrial life must have found its way to emerging New Zealand, by whatever means, became established and subsequently evolved. Or, alternatively, it survived on islands within the general region of Zealandia and subsequently recolonised the newly emerging New Zealand land mass.

Rocks of Miocene age are particularly widespread in New Zealand; more widespread than for any other slice of geological time. Accordingly, Miocene shelly fossils are the most common fossils encountered by New Zealanders, especially in the North Island. What is interesting is that these Miocene rocks are mostly marine sediments. Although New Zealand was emergent and rising, the land area was nevertheless restricted, certainly no more extensive than it is today and probably much less.

Apart from the terrestrial animal and bird fossils described above from Early Miocene sedimentary rocks of central Otago, there are few other significant records of Miocene terrestrial animal life in New Zealand. However, we do have a much better record of plant life, based on fossil leaves, pollen and spores. But the marine fossils are the clear winners: there are literally thousands of localities recorded, and a rich diversity of fossil molluscs, in particular, of bivalves (clams) and gastropods (snails).

17. Evidence of warm subtropical conditions in Northland, 16–11 million years ago in Miocene time. These are fossil coconuts, *Cocos zeylandica*, from well-known fossil plant beds near Coopers Beach. They are all small (less than 5 cm long) and relate to a species of coconut that is now extinct.

18. A large fossil volute gastropod, *Spinomelon parki*, from Lake Waitaki, north Otago, a distant ancestor of modern *Alcithoe*, common around New Zealand's coasts today. This specimen is from rocks of Late Oligocene age, 27–25 million years old. It is named after James Park, professor of mines at the University of Otago early in the twentieth century.

14/ The Rise of the Southern Alps (5–0 Million Years Ago)

This chapter is concerned with the past 5 million years of New Zealand's history: Pliocene and Pleistocene time. Understandably, it is the best-known geological history because it is so young compared with the older record, and it is more reliable. The further back in time we go, the less reliable the record. Much has happened geologically in New Zealand in the past 5 million years. Here we shall address only some of the most significant aspects of this history and focus specifically on what has happened and how we know it has happened.

A Fine Marine Record – Finer Than Most!

Pliocene and Pleistocene marine sediments of New Zealand are especially significant internationally because there are so few places on Earth with such a good record. The reason for this is tectonic uplift; in most other places globally, Pliocene and Pleistocene marine sediments are still under the sea.

In New Zealand, marine sediments of Pliocene–Pleistocene age range make up much of the land in southern North Island and the north-west corner of the South Island. The sediments vary in thickness from a few hundred to thousands of metres. In northern Hawke's Bay, for example, more than 9 kilometres of sediment are preserved.

The source of all this sediment comes mainly from the rising axial ranges of the eastern North Island and the Southern Alps. In other words, it is largely derived from the greywacke basement rock. To put this in perspective, the sediments shed by the erosion of the rising mountain ranges of New Zealand provide an amazing record of environmental change in the New Zealand region during the past 5 million years. The scientific world has also discovered clever ways to make sense of this extraordinary record.

The record is not easy to extract information from. Deciphering is necessary and 'reading rocks' requires specialist knowledge and language. Geologists are required to have an understanding of the rates of natural processes that are at work in the natural environment, and many of these processes are operating over time intervals that are much greater than the life span of an individual human being.

The Rocks of the Wanganui Basin and the East Coast Basin

By far the most important sequences of marine Pliocene–Pleistocene sediments are within the Wanganui Basin on the west side of the southern North Island, and the East Coast Basin on the eastern side. These rocks have been intensely studied and dated because they offer so much insight about processes that govern global change. This can be achieved only because of highly accurate and precise age control, measurable at a resolution of 1000–10,000 years for the past few million. This is an astounding achievement, yet such high resolution is required to make accurate risk assessments of natural hazards such as vulcanism, earthquakes and tsunamis.

In turn, all this newly acquired understanding has been supported by highly detailed knowledge of world climate and sea-level changes during the past 5 million years, a great deal of which is based on painstaking analysis of drill-cores from the deep sea.

Cyclicity

One of the most striking features of these young Pliocene–Pleistocene sediments is the cyclical way in which they occur. This cyclicity is particularly evident in deeply incised river valleys and in road cuttings. The best places to observe this are in the central North Island river valleys, especially those of the Taranaki and Wanganui regions, but also in the Rangitikei River valley. The rocks are layered in rhythms, as if there were a natural pulse, beat or refrain that is endlessly repeated.

What does it all mean? Why are the sediments organised in that way? Is it something to do with tides or currents or something else? To most of us, the organised structure of sedimentary rock is either bewildering and/or unimportant, but geologists have skilfully unravelled the sense of it all. Their explanation may not necessarily be right, but it is very compelling. Sediments are the product of erosion and are normally deposited by flowing fluids, such as water. By studying sediments (sedimentology) and the way in which they are distributed and laid down or deposited, we can learn a great deal about the nature of the fluids involved in their transport and deposition. For example, we can deduce water depth, distance from land, and whether or not sea level was rising or falling.

This raises another question: what controls the rise and fall of sea level?

The Mighty Sun and Earth's Orbit

We now know that what we are observing is a pattern of sedimentation that is governed by global climate change, attributed directly to long-term variations in the Earth's orbit around the Sun. A Serbian scientist, Milutin Milankovitch (1879–1958), was the first to recognise these variations in terms of their significance for global climate change, and hence they are referred to as Milankovitch cycles. The first is eccentricity, the variation in the elliptic shape of the Earth's trajectory around the Sun. Second is obliquity, the variation in the tilt (an oblique angle) of the Earth's spin axis to the plane of its orbit. Third is precession, the wobble of the axis generated as the Earth spins. These three factors determine how much insolation, or heat energy, reaches the Earth from the Sun. This, in turn, determines how cold it is at the Earth's poles.

The approximate periodicities (periods of time) of these three factors are 21,000 years for precession, 41,000 years for obliquity and 100,000 years for eccentricity. Together, these factors have determined predictable global climate change certainly for the past 65 million years, in other words all of Cenozoic time. Since the development of the Antarctic ice sheet about 34 million years ago, the 100,000- and 41,000-year cycles have been stronger; they have been amplified. This means the oscillation in sea level has increased, with sea level dropping further during ice ages or glacials, and rising higher during inter-glacials.

Ice Controls Global Sea Level

Ice accumulates during glacial periods and melts during the warmer interglacial periods. The change in ice volume is responsible for the raising and lowering of the sea level, and this has happened consistently within a range (amplitude) of 100–130 metres during the past 800,000 years. For example, only 20,000 years ago, sea level was 130 metres below its present level.

Global sea level rose by about 3 centimetres from 1997–2007. About half of this rise is attributed to the melting of ice and snow. The other half is attributed to thermal expansion; as things heat up they expand, as does water.

The Amazing Memory of Oxygen

One of the curious effects of freezing water is that it has a selection effect at the atomic level. It selectively freezes the lighter oxygen isotope known as ^{16}O. In terms of chemistry, water is H_2O, in other words a water molecule comprises two atoms of hydrogen to one atom of oxygen. But, as with all elements, there are isotopes. Hydrogen has two isotopes (known as hydrogen and tritium) and oxygen has more, but the common naturally occurring isotopes are ^{16}O and ^{18}O. Isotopes vary in mass (that is, in the number of neutrons within the nucleus). With a glacial period, ice builds up, soaking up ^{16}O, and as a consequence sea water becomes enriched in the heavier isotope ^{18}O.

In short, the isotopic composition of oxygen in sea water can be related to the amount of ice on the surface of the Earth, and this in turn can be related to the amount of solar energy reaching the Earth's surface, and hence to climate.

This chemical phenomenon has proved to be a godsend. Geologists and climatologists have been able to exploit oxygen isotopes and determine a comprehensive history of global climate change.

A History of Sea Water

You might wonder how it is that we are able to determine the history of oxygen isotope composition in sea water. And whether, for instance, there are places where sea water is preserved in rocks.

The answers, once again, all have to do with fossils, and especially fossil plankton. Those single-celled marine organisms that grow skeletons of either silica (SiO_2) or calcium carbonate ($CaCO_3$), absorb the oxygen they need for construction purposes directly from the sea water that they are living in. Their exquisite mineral skeletons therefore retain a record of the oxygen isotope composition of sea water at the time that they grew. The oxygen is locked up securely in a robust, solid

The Rise of the Southern Alps (5—0 Million Years Ago)

Eccentricity
(100,000 years)

Obliquity
(41,000 years)

Precession
(21,000 years)

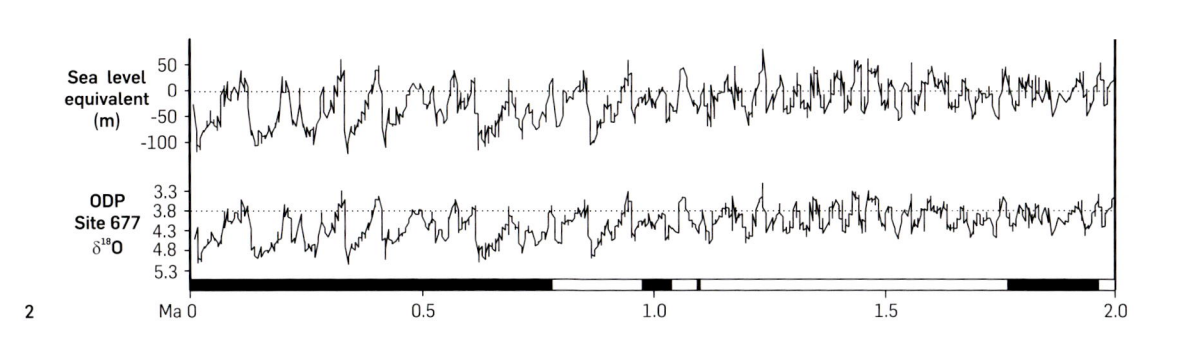

1

2

1. As the Earth orbits around the Sun, it experiences three orbital irregularities: a variation in eccentricity (every 100,000 years), obliquity (every 41,000 years), and precession (21,000 years). These orbital variations determine the distance between the Sun and the Earth and the amount of solar energy that reaches the Earth's surface. This in turn profoundly affects global climate. In time, much of Earth's orbital history will be documented from the study of climate change. The Earth's memory banks will reveal all.

2. Making use of oxygen to unravel the history of the oceans. Oxygen atoms derived from sea water becomes incorporated in shells (as calcite, $CaCO_3$) secreted by marine organisms such as plankton (foraminifera) and shellfish (molluscs and brachiopods). These skeletal structures capture and preserve the isotopic composition of sea water at the time the organism was alive. Through Earth's long history, the isotopic composition of oxygen in sea water has changed because of water temperature. These curves show the astonishingly close match in oxygen isotope ratio ($\delta^{18}O$ is the ratio of ^{18}O to ^{16}O isotopes) as recorded from Ocean Drilling Programme drill-hole Site 677, to sea level variation as measured in height (metres above and below modern-day sea level), for the past 2 million years.

crystalline form and is not subject to subsequent chemical alteration. The oxygen isotope signature is preserved, as if locked up in a time capsule.

Geologists have systematically sampled drill-cores extracted from the sea floor, and have painstakingly extracted fossil plankton from sedimentary rocks. They have then analysed the fossils for their oxygen isotope composition. Since the late 1960s a huge database has been acquired. This is the basis for our detailed knowledge of the history of the oceans, and hence past climate change. The result is known as the global oxygen isotope curve.

Nick Shackleton

The person who pioneered this line of enquiry was Nick Shackleton, a descendent of none other than his namesake of Antarctic exploration fame. He was based at Cambridge University in England and began his research in the late 1960s. Today there are oxygen isotope laboratories all over the world, including several in New Zealand. It is a straightforward process to extract oxygen, particularly from carbonate, and analyse it in a mass spectrometer. It is more difficult to extract oxygen from silica and silicates, but this is also routinely done.

Strontium in Sea Water

However, oxygen is not the only chemical memory available to us. There are others such as strontium, which behaves like calcium in calcium carbonate. The natural variation of strontium composition in sea water through time is also known. The global strontium isotope curve also provides valuable insight into Earth history.

Magnetostratigraphy

Magnetostratigraphy is the record of polarity changes of the Earth's magnetic field through time (this is referred to as the Geomagnetic Polarity Time Scale). During the past 5 million years, there have been 22 changes in polarity. The most recent time the poles flipped was 0.78 million years ago. The duration of a magnetic polarity change event is considerable, but after decades of very thorough investigation, the events of the past 5 million years are extremely well understood, and they all have been named. The present 'normal' event is named the Brunhes Event, as opposed to the previous 'reverse' event, when the North Pole was the South Pole, which is called the Matuyama Event.

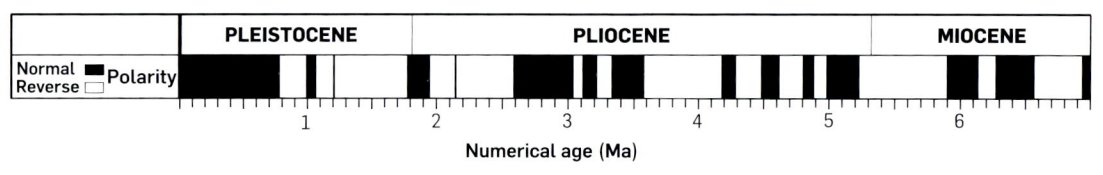

3. Following in the footsteps of Urey and Emiliani, Cambridge scientist Nick Shackleton broke new ground when he perfected techniques for unravelling the history of the oceans. He did so using the precise measurement of oxygen and carbon compositions of calcite (CaCO$_3$) in fossil plankton preserved within marine sediments.

The Rise of the Southern Alps (5–0 Million Years Ago)

4. Making use of the Earth's magnetic memory. The Earth behaves like a huge electromagnetic dynamo, but the polarity of the magnetic field can flip so that north becomes south and south becomes north. The record of the Earth's magnetic polarity (known as the Geomagnetic Polarity Timescale) has become incredibly important to geologists. The timing or ages of the actual changes or flips in polarity are now extremely well calibrated for the past 75 million years. Normal polarity is as it exists today. This segment of timescale shows polarity changes over the past 6 million years. Note how irregular it is and how short-lived some polarity changes have been.

5. This map highlights the volcanic activity in New Zealand during the past 5 million years. Most of it has been in the North Island. Volcanic rocks of Pliocene age (5–1.8 million years ago) are distinguished from Pleistocene volcanic rocks (1.8 million years to present day). The older pre-Miocene rocks of New Zealand are shown in pale blue.

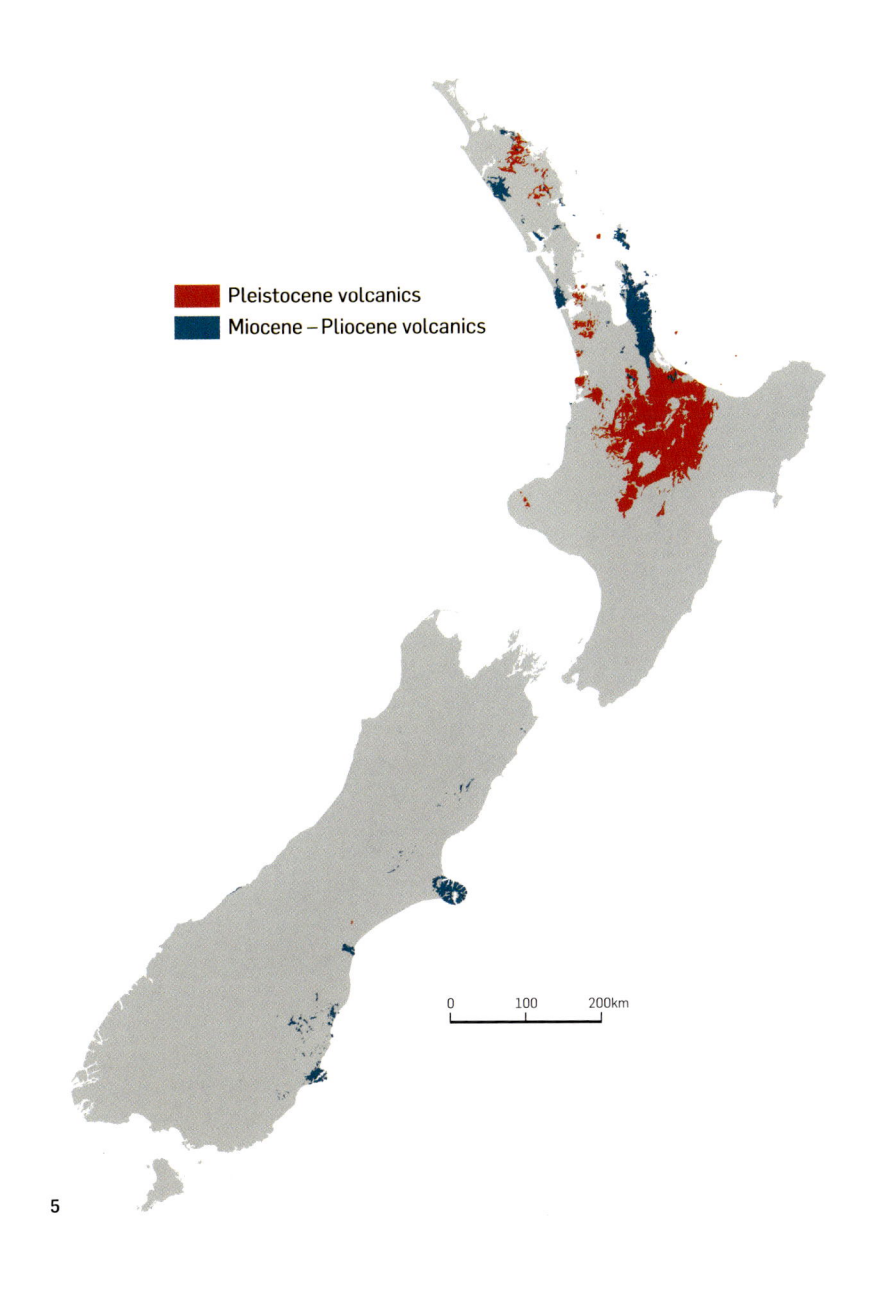

6. The Southern Alps as seen from Banks Peninsula, near Christchurch. Entirely within the leading western edge of the Pacific Plate, they appear as a long breaking wave on the horizon, a great ripple in the Earth's crust, rearing up as a wall of rock immediately adjacent to the collision front with the Australian Plate. This is plate tectonism in action, and it is happening before our eyes.

Iron is the fourth most common element in the Earth's crust and is abundant in most environments in one form or another. Inevitably, it participates in rock-forming processes, including the formation of some mineral cements. Sediments turn into sedimentary rocks with burial, and in the process minerals grow, including iron oxides that form readily from any available iron. These minerals acquire the prevailing magnetic polarity of the Earth's magnetic field. They lock in a memory of the Earth's polarity, normal or reverse, and very little iron is needed to measure this. Using drill-cores of sedimentary rocks, it is a very straightforward process to take samples and determine the polarity of the rock.

Masses of Data: Integrated Stratigraphy

With all this analytical fire-power at our disposal, a remarkably detailed geological history has been figured out from the marine sediments in the Wanganui and East Coast basins for the past 5 million years. The sediments are rich in fossil plankton, shelly fossils and fossil pollen. The fossils provide some age control (biostratigraphy), but, most importantly, they are the source of the oxygen isotope data. The magnetostratigraphy helps calibrate the age of the sediments, as do volcanic ash beds derived from the Taupo Volcanic Zone.

Within these rocks are also preserved fossil soils (paleosols); they are not all marine! Not only is there a remarkable record of climate change, there is also a remarkable record of fluctuating sea level, superimposed on a record of tectonic uplift. It all sounds complicated, and it is. But the integration of all the effects of these processes has produced an enormously satisfying story.

These young rocks of the Wanganui and East Coast basins are locally referred to as 'papa'. They are relatively soft and easily eroded. They seem inconsequential and yet they are of international repute. They are revealed to us by virtue of tectonic uplift associated with plate collision that has been especially vigorous for the past 5 million years.

Now that we have considered some of the North Island, let us consider more effects in the South Island.

The Southern Alps

From space, the most striking feature in the landscape of the South Island is undoubtedly the Alpine Fault. It stands out as a very long, conspicuous lineament, sharp as a knife-cut. It runs south-west to north-east up the West Coast along the western edge of the Southern Alps.

On the ground, however, by far the most striking feature is the Southern Alps themselves. Looking west from the Canterbury Plains they appear as a wall of mountains that form an endless jagged chain along the distant horizon. From the West Coast, they are closer and even more spectacular.

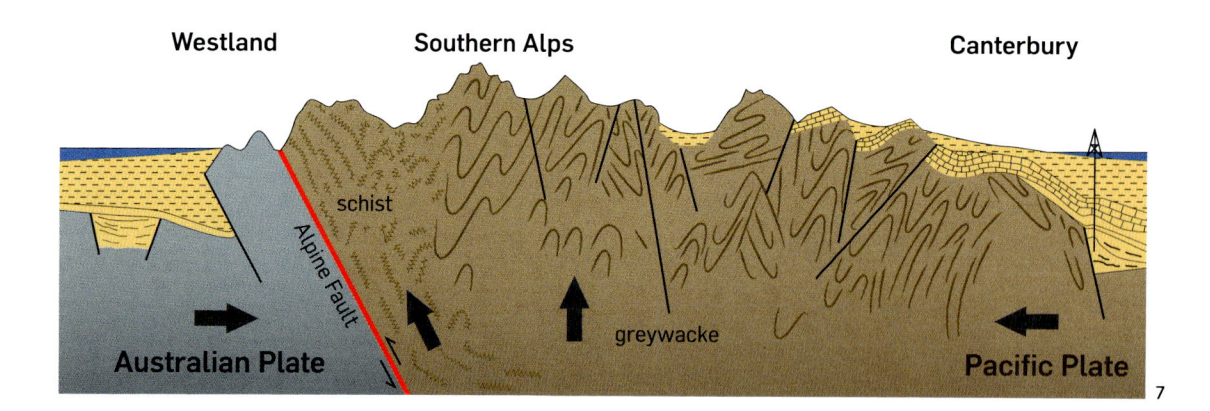

7. To the west of the Southern Alps, older granite and metamorphic rocks collide with younger rocks to the east. Most of these eastern rocks have been brought to the surface as the land has risen within the past few million years. Only remnants of the much younger limestone, mudstone and sandstone cover sediments that once draped a more extensive submarine Zealandia remain. They can today be seen in places like Castle Hill Basin, inland Canterbury.

They seem like an impenetrable barrier. However, although the Southern Alps are long (in excess of 500 kilometres), they are not very wide: only a few tens of kilometres in places.

These mountains constitute a giant leaping wave, a great roller cast in stone, generated by the collision of continental crust on the Australian Plate to the west of the Alpine Fault, and continental crust on the Pacific Plate to the east of the Alpine Fault. It is the eastern side that is rising.

The Southern Alps are directly analogous to the better-known continent–continent collisions that have given rise to the Alps in Europe, and to the Himalayas in Asia. The Southern Alps are a superb example of a surprisingly 'clean' fight between continental crusts, although the Australian Plate is the stronger, holding its own while the Pacific Plate is forced to rise.

Special Characteristics

The Southern Alps are perhaps more elegant and majestic than their namesake in Europe or even the Himalayas. They are certainly less complex and less 'messy' than is the case in Europe or Asia. The main reasons for this are: firstly, that the crust is thinner in New Zealand; secondly, the collision is markedly transpressional (that is, it is not head-to-head but oblique, with twice as much movement sideways as vertical); and thirdly, the rocks on the Pacific Plate are slippery, dominated by schist with an abundance of flat minerals.

There is a fourth reason: the plate boundary collision in New Zealand is fresher, more youthful. Although the Alpine Fault can be traced back to its inception 45 million years ago, it became vigorous about 26 million years ago, and has been especially active in the past 5 million years.

The fifth and final reason for their relative lack of complexity is that the Southern Alps have not been as severely eroded by glaciers as the European Alps and the Himalayas. The ice ages of the past 2 million years were kinder to New Zealand than Europe and Asia. This 'freshness' makes the Southern Alps seem like a mere youth compared with the more senile but more venerated collisions in Europe and Asia.

We know all this from reading the rocks. The faster the

mountains are rising, the more rapid the erosion, and the greater the sediment accumulation on the sea floor adjacent to the New Zealand land mass.

The Crash Zone

The collision zone is interesting and has been much studied. It resembles the highly crumpled front end of a simulated car crash, whereby the front end of the car is smashed, while the back remains intact as the car races forward and slows on impact. It sounds far-fetched, especially when you consider that the rate of impact is measurable in millimetres per year, yet this analogy is an accurate one.

The collision is causing the crust to pile up both at depth and above. The crust is being compressed. A very substantial 'root' is being generated and this effectively thickens the crust, creating an increased localised buoyancy effect parallel to the plate margin and all along the length of continent–continent contact. This buoyancy produces the mountains.

The rate of collision is measurable and can be expressed in terms of movement on the Alpine Fault. Most of the motion is sideways: it is moving 20–30 millimetres per year horizontally. The Southern Alps are also moving vertically. So, relative to the lower western side of the Alpine Fault, the eastern mountainous side is rising by up to 10 millimetres per year.

From knowing the rate of plate motion, and the history of sea-floor spreading, we are able to deduce that the plate boundary collision has resulted in shortening or compressing of the New Zealand crust by about 60 kilometres. In other words, a length of the Earth's crust has been shortened by a distance of 60 kilometres. Most of this crumpling and squeezing has been taken up within that frontal collision zone that is defined by the Southern Alps.

Furthermore, we can deduce that the Southern Alps represent the remains of more than 20 kilometres of uplifted crust. Had the crust not been eroded away almost as fast as it rose, the Southern Alps would stand another 20,000 metres above their present height. In a sense, the Southern Alps are a mere shadow of what they might have been. It has been estimated that about half a million cubic kilometres of rock has been eroded away. And it has all gone into the sea, as sediment.

The rise of the Southern Alps has been, and is, the most profound and most conspicuous product of the collision between the Australian and Pacific plates. It is this continent–continent collision that has given rise to the South Island of New Zealand as we know it. The Southern Alps mark the locus of greatest action in the collision. Yet everything else has been dragged up with it.

The length of the collision front is determined by the extent of available continental crust. This is the reason why the South Island is as long as it is. It cannot be any longer because there is no more continental crust on the Australian Plate south of Milford Sound, and not much on the Pacific Plate south of Fiordland.

Folding the Land

The Southern Alps are the most dramatic part of the 'crash zone'. Further eastwards the effects diminish, but are nevertheless still strong. The basin and range topography of east Otago reflects large-scale warping or folding. Smaller-scale but similar folding is happening in the Halcombe area near Fielding in the North Island. Halcombe Hill is a growing fold.

The southern part of the North Island is a crumple zone, too. If you squeeze a sheet of corrugated cardboard, the valleys go down and the ridges go up. This is more or less what is happening to the Earth's crust, and is especially evident in the Wellington and Wairarapa areas. The only difference is that the crustal forces are transpressional, so the lateral or sideways motion is more than twice the vertical motion. Hence, during the 1855 earthquake, the incredibly large 17-metre dislocation on the Wairarapa Fault, yet only 6.5 metres of vertical motion. The corrugated cardboard analogy helps explain Lake Wairarapa and the Hutt Valley: they are both being driven down, whereas the ranges to either side are being driven up.

8.

9.

8. Compelling geological evidence shows that up until about 2 million years ago the Chatham Islands were totally submerged. After millions of years of submergence, they have been pushed up out of the sea by tectonic forces. The landscape in northern Chatham Island is eerie in that it really can be regarded as a dried-out submarine seascape, complete with upstanding conical-shaped eroded remnants of submarine volcanoes. Within the past 50,000 to 100,000 years, peat has formed and has smoothed out the gentle subdued topography, enhancing its smoothed flatness.

9. The Bounty Islands are other bits of Zealandia with their heads above water. Like Stewart Island and The Snares, this is the remnant of an upstanding body of granite. Now almost gone, the surface area of the Bounty Islands would have been much greater than it is now.

Landslides
Landslides play a significant role in shaping our mountain landscape. It is now clear to us that you cannot poke large masses of rock up in the air and expect them to stay there forever. Given the right conditions, they will fall under the influence of gravity. They do not necessarily fall in response to any particular trigger. And they can fall without warning. On the basis of a number of historic events, as well as detailed research by geologists at GNS Science, we are now certain that landslides are largely responsible for much of the shapeliness of our mountains.

An example of this was the spectacular collapse of the summit of Aoraki/Mt Cook on 14 December 1991. New Zealand's highest peak lost almost 10 metres of its stature in a matter of seconds as 12–14 million cubic metres of rock plummeted downwards onto the Tasman Glacier and carried with it at least another 40 million cubic metres of rock and ice. The impact unleashed an earthquake of magnitude 3.9. Yet the landslide was not triggered by any obvious cause, such as heavy rain or an earthquake. It happened on a quiet, clear, starlit night.

More recently, a helicopter pilot reported a large landslide in a remote part of Fiordland. This large rock fall is located to the east of the Olivine Ice Plateau some 25 kilometres south-west of Mt Aspiring and 47 kilometres north-east of Milford Sound in the Mt Aspiring National Park. Forensic sleuthing by landslide experts at GNS Science has since established that an estimated 1.5 million cubic metres collapsed from about 600 metres above the John Inglis Valley on 12 December 2006. Two small earthquakes were associated with this fall, one 1.8 in magnitude, and the other 2.9.

Onset of Continental Rifting in the North Island
It was during Pliocene time that the Taupo Volcanic Zone began to rift. The result is the spectacular 'Y-shape' of modern-day North Island, with northern North Island projecting in a north-west direction and the eastern coast of the southern North Island up to East Cape projecting in a north-east direction. These two great projections were once more or less parallel. The yawning gape is the product of continental rifting, often referred to as 'back-arc rifting' because it is considered to be 'behind' the volcanic arc.

In this case, the volcanic arc includes Ruapehu to the south and stretches away to the north-east along a line through White Island and on up to the Kermadec Islands. This is the northern New Zealand segment of the classic Pacific Ring of Fire. A southern New Zealand segment continues south of the Alpine Fault along a line from Solander Island to Macquarie Island.

The rift has effectively rotated Northland away from East Cape, and it is still happening. Direct measurements indicate an east–west motion of about 10 millimetres per year. It is this stretching process that is largely responsible for the huge rhyolite calderas in the Taupo Volcanic Zone. They are a manifestation of fresh continental crust forming at depth, producing new granite (fresh cream).

The rift is also responsible for making Auckland relatively safe from earthquake hazards. The rifting has removed the Auckland area ever further away from the plate boundary. However, it is still subject to some earthquakes associated with active normal faults along the western margin of the rift zone at the northern end of the Hauraki Gulf. These faults are under tension and permit small-scale stretching of the crust. This is what happened during the 1987 Edgecumbe Earthquake, but it related to the eastern margin of the rift zone.

The Rise and Emergence of the Chatham Islands
A number of relatively small volcanoes erupted 5–4 million years ago, at the eastern end of the Chatham Rise in what is nowadays the Chatham Islands. These eruptions were comparable to the Auckland volcanoes, in that they were all basalt. But what makes them interesting is that they are all submarine and that there were lots of them. The Chatham Islands sport the remains of at least 20 distinct volcanoes of this age. Of these, the biggest form prominent hills or high points in the landscape, such as Maunganui, The Sisters, Cape Young, Hikurangi, Korako, Motuporoporo,

10. A botanical oddity, the vegetable sheep is actually a member of the daisy family, with one sheep-like clump composed of thousands of tiny plants. Like most of New Zealand's alpine plants, it dates back just several million years to the time when mountain uplift began to accelerate.

11. One of New Zealand's most spectacular fossil localities: fossil scallop beds at Momoe-a-toa, northern coast of Chatham Island. These fossils are 5 million years old and represent a sea floor, at least 100 m deep. The collectors here are all professional paleontologists: Daphne Lee, Alan Beu and Phillip Maxwell.

12. A fossil shellbed of scallops belonging to the extinct species *Phialopecten marwicki*, about five million years old, of Early Pliocene age. Fossil scallops preserve extremely well because of their robust calcite shells, and are widespread in marine sediments of Oligocene to Pleistocene age. Some species of scallop such as *Zygochlamys delicatula* (Queen Scallop) are excellent indicators of water temperature. Alan Beu is an authority on fossil scallops and has demonstrated how useful this species is as an indicator of change in sea-water temperature during the glaciations of the past 2 million years.

11

Motuariki, Hokopoi, Dieffenbach, Matakitaki, Star Keys, Southeast Island and probably the Murumurus.

Most significantly, submarine volcanic rocks of this age range from the highest topography within the archipelago in the south-west corner of Chatham Island. Note that the highest point is still less than 300 metres above sea level. However, this nevertheless implies that, but for the upstanding Mangere Volcano (6 million years old), the entire Chatham Islands region was totally submerged 4 million years ago. With time, even Mangere became submerged. Then, about 2 million years ago, in Early Pliocene time, land reappeared. The reason for this can only be tectonic uplift and there has been land ever since. The exact mechanism or process of uplift has yet to be determined, but it probably involves a combination of one or more of three possible factors: 1) far-field tectonic east–west buckling of the continental crust forming the Chatham Rise as a consequence of plate boundary collision through mainland New Zealand; 2) localised inflation of the crust in the region of the Chatham Islands due to magmatic processes at depth, as indicated by the relative abundance of volcanic eruption centres in the area during the past 6 million years; and/or 3) a broader scale inflation of the crust due to thermal up-welling within the mantle. Future geophysical research should be able to determine exactly what the uplift process is.

The Establishment of New Zealand Biota

By the end of Miocene time 5 million years ago, we assume that all the forebears of the modern New Zealand biota were well established. However, if we were to observe strictly what the fossil record tells us, then much of the New Zealand biota would be pronounced absent until Pleistocene time, 1.8 million years ago.

The record is remarkably thin, particularly for our larger iconic animals and birds. The plant pollen record is understandably better, but what it tells us is that most of it, if not all of it, relates to a vegetation that is less than 23 million years old. Modern research, using molecular biology, is exploring this idea and testing it, plant group by plant group, and animal group by animal group.

13. This spectacular shore platform and ledged headland is Motutapu Point, the very north-west corner of Pitt Island in the Chatham Islands. This is one of the finest fossil localities in New Zealand with exceptionally rich and diverse Pliocene shelly fossils. More than 350 species of fossil have been recorded from this one locality.

14. A fossil paua, *Haliotis* (*Notohaliotis*), from rocks of Oligocene age, 30–25 million years old, inland of Kaikoura. They are surprisingly rare in the fossil record, partly because they occur close to the shore where waves erode their traces, and partly because the shells are largely comprised of aragonite, which preserves much less easily than calcite.

15. A fossil flounder perhaps? This fossil flat-fish (referred to as *Pleuronectiformes*) was found during construction (using explosives) of a garage floor on a farm near Te Pohue, inland Hawke's Bay. So many of New Zealand's most interesting fossils have been spotted by ordinary New Zealanders. This specimen was within volcanic sandstone of Pliocene age, about 3 million years old.

16. Of the 27 species of native freshwater fish in New Zealand, 13 belong to a family referred to as the galaxiids. One of the more common species is the banded kokopu (*Galaxias fasciatus*). Like their galaxiid cousins, they are small and tend to be secretive and nocturnal. Scientists are questioning how easy it would it have been for these freshwater species to have colonised New Zealand if it had been submerged 23 million years ago.

17. A famous seasonal component of New Zealand cuisine is whitebait. These tiny delicacies are the juveniles of five species of galaxiid fish. Their lifecycle is complex, part freshwater, part marine and they swarm in their millions in spring (September to November). Is it possible that, once conditions were to their liking, these freshwater fish colonised New Zealand from another landmass such as Australia over the past 20 or so million years?

18. Fossils of birds that are older than 1.8 million years old (in other words of pre-Pleistocene age) are rare in New Zealand. However, fossils of this unusual 'false toothed' bird have been found in Pliocene rocks from Canterbury and Taranaki. The fossil (A) is referred to as *Pseudodontornis*, a bony-toothed pelican (or pelagonorthid), and the actual bird (B) may have looked like this reconstruction on display in Puke Ariki Museum, New Plymouth.

Pliocene Marine Fossils

Once again, the marine record is much more comprehensive and informative than the terrestrial record. There are several rich and diverse Pliocene fossil shellbed localities in Northland and the Chatham Islands. They comprise more than 350 species of mollusc alone! So our understanding of life in the New Zealand seas is reasonably well understood. Modern analysis indicates that the fossil record in general for Cenozoic molluscs is about 40 percent representative of the total molluscan fauna.

Alan Beu (GNS Science) is a leading authority on the Pliocene and Pleistocene fossil molluscs of New Zealand. Together with Phillip Maxwell and artist Ron Brazier, they have written the finest compendium available on the Cenozoic molluscs of New Zealand, and thereby Zealandia!

Among his many achievements, Alan Beu has made sense of the fossil scallops and pauas, classic icons of New Zealand. He has also been able to show the close relationship between certain species of scallop and water temperature. Using this information, it has been possible to map the changing distribution of *Chlamys delicatula* in time and space for the past 2 million years. This record serves as an accurate proxy for sea temperature, and hence climate change. Furthermore, it has been confirmed by oxygen isotope studies.

Fossil Flounder or Patiki

Fossil fish are not common. It is little wonder they are only rarely preserved and found as fossils, when you consider how fragile fish bones are. However, in the 1970s, the Kings, a farming family near Pohue on the Napier–Taupo Road, were excavating in tough Pliocene marine sedimentary rock for a new garage floor using explosives. In the blast, a block flew open to reveal an almost perfectly preserved flat fish, comparable in size to a small flounder.

The fossil is preserved on two rock surfaces, just as all fossils must be if they are preserved in rock. There is always a right side and a left side, and, of course, they fit neatly together like a hand and glove. In this instance, one half ended up with the local teacher and the other half is now part of the National Paleontology Collection with GNS Science. For the fish to be preserved, it must have been overwhelmed and rapidly buried by sediment. This particular fossil is about 4 million years old, and no other fossil quite like it is known from elsewhere in New Zealand. It is unique and of great interest because of its biological significance as an ancestral flounder, perhaps. It is as yet unstudied and undescribed.

This is one of those rare finds that is worth knowing about, a real treasure. There must be thousands of stories like this around New Zealand, especially out there in the countryside. Each one has the capacity to add to our scientific understanding of the natural heritage of New Zealand and consequently add richness to our lives.

Another Fishing Story

In more recent times, there has been another great fish find, this time in lake sediments near Ormond, inland from Gisborne. Liz Kennedy was collecting for plant fossils and by chance discovered a superbly preserved fish that has been identified by Bob McDowall (NIWA) as the first fossil grayling to be recorded from New Zealand. The grayling was a significant native freshwater fish that became extinct in the 1920s.

The Toothless Bony-Toothed Pelicans

Some of the more interesting fossil discoveries from Miocene and Pliocene rocks are the 'false-toothed' birds or 'pseudodontorns', now placed in the family Pelagornithidae. These were very large coastal seabirds akin to pelicans and gannets that looked menacing with their serrated, tooth-like projections along the side of their bill. A spectacular reconstruction of what this bird must have looked like is on display at Puke Ariki Museum in New Plymouth. It is based on fossils found in Miocene and Pliocene rocks in north Canterbury and near Hawera in Taranaki. To see this is quite sobering. It informs us of just how different life in New Zealand could have been in Miocene and Pliocene time, and makes us appreciate the slender basis of our knowledge: just a few fragmental fossils. But without them, what would we know? Fossils of bony-toothed pelicans have been recognised from three fossil localities in New Zealand. The fossils amount to a partial skull and four limb bones in total. It isn't much, but it's enough.

18A

18B

15/ Riding the Past 120,000 Years

This chapter considers the history of New Zealand over the past 120,000 years. This time span might seem trivial compared with the many millions of years that we have considered so far, and yet it is still an enormous chunk of time.

To put this time frame in perspective, it is still much longer than 50,000 years. This is a magic number in the geological world because radiocarbon dating

1. Evidence of a relatively recent forest can be seen here on the beach at Titahi Bay, north of Wellington city. Tree stumps and a paleosol (fossil soil), a clay-rich, black mud riddled with matted roots and forest floor litter, appear intermittently through the sand, particularly during winter storms. Dating from the most recent interglacial period 120,000 years ago, the fossil forest floor preserved here is similar to others such as at Ocean Beach, Wairarapa. Some tree stumps like this one here with Niamh Campbell, have been identified as totara. Others are matai.

techniques can be applied only to materials that are less than 50,000 years old. Other dating techniques are required for material that is older than this. In other words, radiocarbon-dating is not applicable to most geological problems because they generally involve rocks and fossils that are much older than 50,000 years.

The past 120,000 years takes in some of the most significant moments in our history, not the least of which is the arrival of human beings in New Zealand, sometime between AD 1250 and 1300. However, let us first consider the pertinent geological events prior to our arrival, namely the last interglacial and glacial period.

The Last Interglacial

The last time global sea level was as high as it is today was about 120,000 years ago, hence the choice of this time span for this chapter. This moment in time marks the warmest moment of the past interglacial period. From then on, the Earth cooled, reaching its coldest 20,000 years ago. This moment is referred to as the 'past glacial'. Since then the Earth has been getting warmer and has been in an interglacial period. It is only a matter of time before the Earth will start cooling again and will head into another glacial period. The sea level has been remarkably stable for the past 6000 years, but rather than falling it has started rising in the last few decades.

2

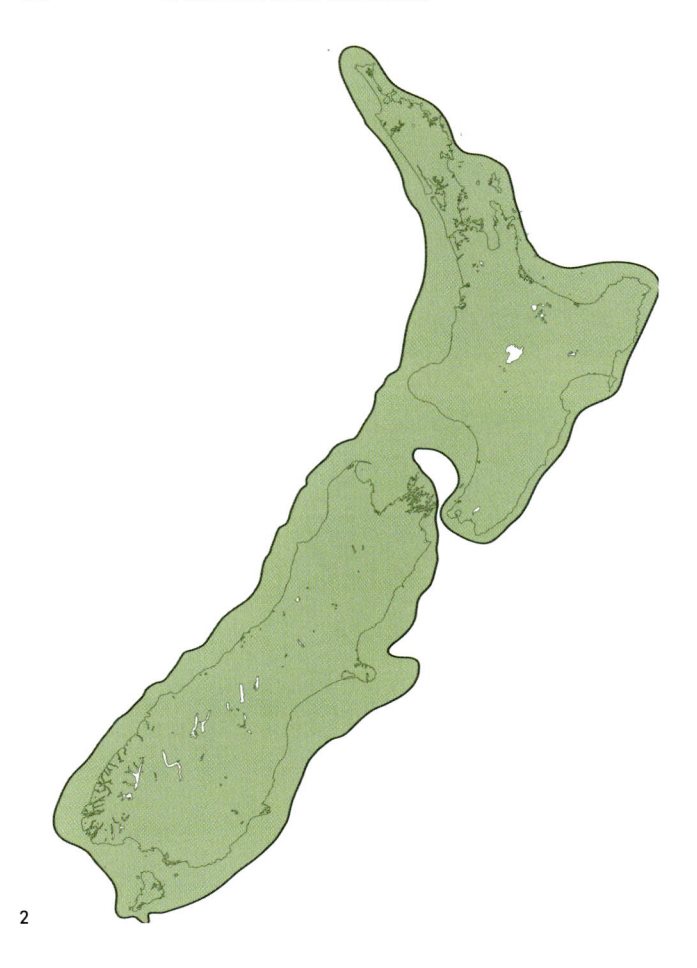

The reasons for this rise are vigorously debated, and the core question is whether the rise can be directly attributed to human activities.

The sea level has been lower than it is today for the past 120,000 years. The climate has also been cooler than today for most of this period. The last time that climate was as warm and as equable as it is now was 120,000 years ago. This might be hard for us to appreciate, but it means that life in New Zealand has not been better than it is right now for at least the past 100,000 years!

It is relatively easy to visualise New Zealand 120,000 years ago because the sea level was more or less at its present level. It left its mark in places. We can detect and identify the old shoreline at that time, especially in those parts of New Zealand where there has been especially rapid uplift, elevating the old 120,000-year coast well above the present sea level.

Titahi Bay Fossil Forest Floor

From fossil plants and pollen, we also know that the vegetation was much the same as today. Fossil forest floors are preserved in places, along with fossil soil and tree stumps. One such locality is at Titahi Bay, just north of Wellington. The remains of a coastal forest are preserved on the beach, including patches of a black, muddy paleosol (ancient soil) about half a metre thick, held together by a mass of rootlets and associated forest litter. There are also tree stumps, mainly of totara and rimu, which is exactly what might be expected.

Although only intermittently seen and primarily at times of the year when the tides and currents have swept the beach clean, this preserved forest is a striking example of a forest floor dating back to the past interglacial age 120,000 years ago.

2. In a glaciation, sea level drops. It does so because so much of the world's water is locked up as ice in the most frigid parts of the Earth: mountainous and polar regions. During the past glacial period, about 20,000 years ago, sea level was about 125 m below its present level. This paleogeographic map shows the extent of the New Zealand land surface area at that time. It was much greater than at present and there was no Cook Strait. Yet this dramatic effect would only have required an average surface temperature drop of 5°C.

3. Today's climate is warm by comparison to most of the past 120,000 years. Glaciers such as the Tasman Glacier would have blanketed the landscape to a much greater degree than now, when all the Earth's rivers of ice are on the retreat.

This is just one of a number of examples recorded from around New Zealand. For instance, at Ocean Beach on the Wairarapa coast of Palliser Bay, a number of conspicuous tree stumps of kahikatea are preserved. These, too, relate to the past interglacial period.

The Past Glaciation

Since 120,000 years ago there have been at least five cooling phases that were more or less governed by the 21,000-year cyclicity of the Earth's precession, the wobble of the spin axis. Each of these cooling phases was accompanied by an ice advance and was colder than the previous one, culminating in the past glacial period or glaciation 20,000 years ago. At that time, average surface temperature globally was about 5°C less than it is today, and the sea level was 125–130 metres below present.

The Glacial Landscape

During the period of the past glaciation, New Zealand would have looked very different from today. Cook Strait would have been a gulf. There was no through-going strait. The North and South islands were one land area, considerably larger in extent than they are today.

Much of the Southern Alps would have been buried under a substantial ice sheet. Much of the present-day high land would have been alpine. The greater part of New Zealand would have been grassland and shrubland, as opposed to forest. Forested areas would have been scattered in protected coastal lowlands, with continuous forest restricted to the northern North Island.

There may have been a few small glaciers in the Tararua Range, but hard evidence is wanting, despite much research. Glaciers would have certainly formed on the major North Island volcanic cones of Ruapehu, Tongariro and Taranaki, but subsequent climatic warming and volcanic eruptions have erased the geological memory of any glaciers. There is no evidence.

Ice Carving

Meanwhile, the glaciers of the South Island were digging in and utterly transformed the landscape. Not only did they sculpt the mountains, they also shunted vast amounts of rock debris to the fringing lowland margins of the mountains and created spaces for the beautiful lakes of the South Island. Glacial moraines and outwash gravels have in turn been redistributed by the many large rivers draining the Alps. Extensive alluvial plains have developed of which the Canterbury Plains are the most spectacular example.

Sand Dunes

New Zealand would have been cold and windy: witness the extent of wind-blown sediments in the New Zealand landscape. These are not necessarily the sole product of the last glacial period, but by virtue of their mobility and instability they do not last in the environment for long. So it is reasonable to assume that the majority of what is preserved is derived from the past glacial period. These sediments include sand dunes and loess (wind-blown dust).

There are very significant areas of extensive sand dunes in that great sweep of country from Levin to Wanganui, and they extend well inland. State Highway 1 cuts right through them, especially between Foxton and Sanson. They form widely spaced, low, elongate ridges snaking across the landscape.

Wind Sculptures

Wind-sculptured stones or ventifacts are sometimes associated with ancient dune sands. Perhaps the best occurrence of ancient ventifacts is in the Waitotara area just north of Wanganui. Here, pebbles and cobbles have been cut and polished by the impact of wind-blown sediment, namely sand and silt. The stones are faceted, beautifully shaped and their surface is lustrous from polishing and buffing. They make spectacular natural sculptures and are testimony to strong relentless wind conditions over sustained periods of time.

Ventifacts are forming today, primarily on exposed and expansive sandy beaches where there is no shortage of sediment. Needless to say, these selected spots in the New Zealand landscape are among the windiest places in the country.

Loess: Dust from a Retreating Sea

Loess is finer-grained than sand and it forms sheets or layers, best seen in road cuttings. (Loess comes from the Swiss German word *lösch*, meaning 'loose'.) This fine sediment or dust is derived mainly from glacial action and is comprised of silt and clay-sized particles of rock that have been mechanically ground by the action of ice, rather like how flour is ground from grains. However, the particles of loess ('rock flour') are recycled and wind-blown. Glacial melt transports the particles down river systems to the sea, where they are deposited on the sea floor. When the sea retreats, winds blow this sediment back onto the land as loess. However, we now know that the majority of loess is derived from exposed river deposits on aluvial plains.

Substantial parts of the New Zealand landscape less than about 200 metres above sea level are mantled in loess deposits, most notably the Timaru area of south Canterbury, where the loess is measurable in metres, and in some places approaches 20 metres in thickness. Loess has a characteristic yellowish colour and forms the massive 'clay' layer that is so widespread beneath the otherwise thin productive soil. It is evident throughout the eastern South Island and Wellington regions in particular.

From a fossil perspective, the loess is very

4. Sand dunes dominate the landscape over a huge area of the western North Island, from the Kapiti Coast to south Taranaki, north of Wanganui. The main highway traverses some of this country between Paekakariki and Bulls, cutting its way across ancient late Pleistocene dune systems. This view is looking inland from near Foxton.

Most of these dunes would have been active during the past glaciation.

5. Testimony to climate change in Pleistocene time, loess is conspicuous all over eastern New Zealand and can be traced as much as 50 km inland, especially in the South Island. This distinctive pale-coloured silt and clay was deposited by wind at times when climate was much cooler and windier than it is today. Layers of loess can be observed in many places, each one relating to a different glacial period during the past 2 million years. Here, near Caroline Bay at Timaru, several loess formations with darker coloured fossil soils (paleosols) between them overlie columnar-jointed Timaru Basalt, the youngest lava flow in the South Island. This is one of several flows that erupted from Mt Horrible, 15 km inland, 2.1 million years ago.

6. Scientific sleuths on the trail of recent geological and climate changes. GNS Science and NIWA scientists have recovered sediment samples from below the bed of Lake Tutira in Hawke's Bay. These 'windows to the past' have revealed some of the turbulent history of the region surrounding the lake that was formed 7400 years ago, including a period about 2000 years ago when 118 major storms swept through over a 181-year time span. They also record ash deposits (tephra) from eruptions within the Taupo Volcanic Zone. This 1 m core, on display at Te Papa, has two tephra layers.

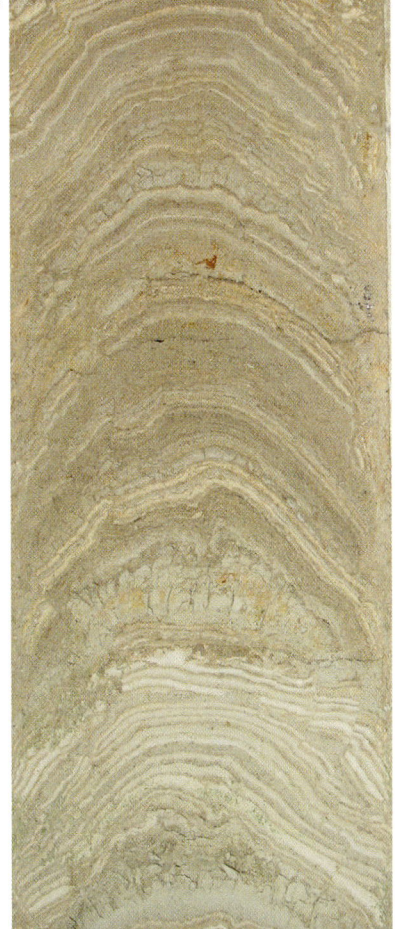

unproductive. There is no doubt that some of the loess is derived from fossil debris such as shells, sponges and plankton, but there are few fossils preserved in the loess. Presumably a dust storm is not catastrophic or voluminous enough to trap and preserve. Nevertheless, some moa bones have been recovered from loess.

Although loess can be discussed as a single entity, it is derived from numerous wind events over long periods of time. Many loess deposits have been recognised and mapped. The persistent dust storms that sometimes strike the Mackenzie Basin and broad-braided river valleys of inland Canterbury, such as the Rakaia and Rangitata, are like loess in the making. Much of the dust is in fact 'rock flour' dumped by rivers draining the Southern Alps.

Climate Change and Flora

Following the glacial maximum 20,000 years ago, the climate slowly improved. From about 14,000 years ago improvement accelerated, and did so again from about 10,000 years ago. We know this from detailed analysis of fossil pollen records. Numerous studies of drill-cores from soil, peat, swamp and lake deposits from all over New Zealand tell the same story of systematic floral change responding rapidly to stepwise climate change.

Drilling into floors of lakes is likely to become a major pastime of New Zealand scientists for years to come, as drill-cores offer by far the best and easiest source of detailed information about ancient New Zealand. One exceptionally well-studied lake is Lake Tutira near Napier. It reveals a superb sedimentary record or history of Late Pleistocene time, with a huge number of 'events' recorded, such as storms, floods and volcanic eruptions superimposed against a background of 'normal' lake life.

The Antiquity of the Modern Coast

Between 20,000 and 6000 years ago, sea level rose by at least 125 metres, and it has been more or less stable ever since. This has profound implications. It means that our modern harbours and estuaries are all very young, at least geologically. They are no older than 20,000 years and no younger than 6000 years. Yet, when we observe them, it is easy to feel that they, and all their associated life, have been there forever.

In terms of human presence, 6000 years is almost like a magic number. If we think of sea level as an edge, then the sea edge is like a razor against the land. When sea level rises, the sea edge cuts and erodes. No wonder there is such a poor record of our human ancestry over this period of time. As coast-dwelling creatures, humans have always lived by the sea, and as the sea rises, we move with it, up slope. It follows that any remains of coastal human occupation from 20,000–6000 years ago will have been destroyed by the sea.

Life in Pre-Human New Zealand

Typically for New Zealand, there is a good plant record but the animal record is sparse. Nevertheless, it is much better than for the previous 15 million years.

A number of very substantial sand dune, swamp and cave deposits are known, and detailed excavation and analysis has established a wealth of information about life in middle to late Pleistocene New Zealand. Most of what we know is based on these few key fossil localities.

The results of many decades of research have established that until humans arrived, there was a complete absence of terrestrial mammals other than bats and seals, and a complete absence of snakes and crocodiles. This was a land of birds, insects, a diversity of lizards and frogs, and the tuatara. It was a very different world from almost anywhere else on Earth.

Te Ana-a-Moe

One example of a locality with substantial deposits of this kind is a single-chamber, pitfall trap cave on the western shore of Te Whanga Lagoon on Chatham Island. This cave is near Te Ana-a-Moe (cave of sleep), which is famous for its petroglyphs, and is a natural dissolution cavern formed within the base of an upstanding bluff of relatively soft limestone of Late Eocene age (37–34 million years old).

The cave must have had an easily accessed, slot-like entrance that attracted birds, but once inside they could not get out. The sides must have been too steep to negotiate and the entrance must have been high above the floor of the cave. Eventually they would die, and their remains accumulated over some period of time.

Radiocarbon dating suggests that they accumulated over a period of about 3000 years, 4000–1000 years ago. This extraordinary deposit was excavated by professional ornithologists and has revealed much of what is known of the prehistoric bird life of the Chatham Islands. The abundance of bird bones recovered is quite staggering and includes at least 15 terrestrial species and eight marine species. No doubt there are other places like this in the extensive limestone terrane of Chatham Island.

Since the arrival of humans in the Chathams between about AD 1400 and 1450 at least 50 percent of the total bird fauna has become extinct.

Southeast Island

What would New Zealand have been like before the arrival of humans? It would have been teeming with bird life, not unlike the sort of ecosystem you can observe today on Kapiti Island. This is the largest of a growing number of offshore islands within New Zealand that is now free of mammalian pests.

Southeast Island in the Chatham Islands is another example. This is a very special place, a jewel in New Zealand's conservation crown. Not only is it home to millions of burrowing sea birds, primarily white-faced storm petrels and sooty shearwaters, it is also home to a number of other extremely rare forest and shore birds. It claims to have, on one small island, more species of rare and endangered birds than anywhere else globally. To visit Southeast Island is a very memorable experience indeed.

The birds aside, the island is highly instructive geologically. Here is a

7. The iconic kiwi, like many of the birds that are descendants of species that have inhabited New Zealand for millions of years, is under threat from predators and loss of habitat.

8. Sometimes described as a 'living fossil', the caterpillar-like *Peripatus* can trace its ancestry back 550 million years, an extraordinarily lengthy time for an order of animals to survive.

9. Only 218 hectares in size, Southeast Island in the Chatham Islands is a biological treasure house with all the bird species recorded by ornithologists in 1871 still present. This is not coincidence: there are no rats on the island. This is a DOC reserve and it is home to four of the world's rarest birds: the Black Robin, Chatham Island Snipe, the Shore Plover and the Chatham Island Petrel.

10

glimpse of what burrowing sea birds can do to a landscape, and a forested landscape at that. They are serious biological excavators, each producing a burrow that is about 2 metres long. They fish all day long and come home late at night, find their way by smell through the forest to their own burrow, and then head off fishing again before dawn. The ground on Southeast Island is treacherous. Apart from the rock platform on the coast, there is nowhere to tread for fear of crashing through the surface and into a burrow!

It is not hard to imagine tracts of mainland New Zealand looking like Southeast Island. In the absence of predators other than birds, why not? Imagine vast populations of burrowing sea birds all around New Zealand. But being predator-free is only one constraining factor. Perhaps just as significant would be the availability of food at the ocean surface.

Vulcanism and Tectonism

Whereas the landscape in the South Island was greatly modified by glaciation during the past 120,000 years, as well as sustained tectonism, in the North Island the most dramatic effects on the landscape have undoubtedly been vulcanism and tectonic uplift. For instance, from Taupo alone, there have probably been at least 120 eruptions based on an average of one every 1000 years. The Wairarapa Fault has also moved probably 120 times, based on the same rate.

Most of the volcanoes in the Auckland area have erupted within the past 120,000 years. Modern research on the Orakei Basin has established at least 90 separate eruptions within the past 90,000 years. This is based on detailed analysis of drill-cores.

The most recent 28 eruptions of Taupo, all within the past 26,500 years, have had a pronounced smoothing effect on the landscape of the central North Island. Vast amounts of pumice and tephra have been dumped in successive eruptions, infilling pre-existing topography.

These eruptions, although of varying size and destructive capacity, can all be dated, and they all have their own physical and chemical characteristics. They are superb time markers in our environment. They all have a biological signature as well, because they mowed down and buried soils and forests.

Some of these effects have had a profound influence on the drainage history of major rivers. The lakes and river valleys of today are not necessarily what they were just a few thousand years ago. For instance, the Waikato River used to flow into the Firth of Thames, but volcanic and tectonic events less than

Rangitoto Island Man and Dog

Since the arrival of man between AD 1250 and 1300 there have been only two eruptions of a rhyolite volcano, namely Tarawera, in about AD 1300 and in 1886 (the latter was a basaltic eruption). There have been successive eruptions of the subduction-related arc volcanoes White Island, Tongariro, Ruapehu and Taranaki. Taranaki last erupted in 1755, and there has been only one eruption of an Auckland volcano – Rangitoto – just 600 years ago.

Human footprints along with dog footprints are preserved in volcanic mud on Motutapu Island adjacent to Rangitoto. This is the clearest possible evidence of the presence of people when Rangitoto erupted, and the volcanic tephra is easily dated.

Unfortunately, when humans arrived in New Zealand, they brought the Polynesian rat, kiore. This single animal species has undoubtedly had more impact on New Zealand wildlife than any other organism, apart from humans. Together they have done the damage. But the spread of kiore has not been confined to New Zealand. It has quietly and systematically altered the face of the Pacific forever.

The First Geologists from Polynesia

In no time at all, the first humans to reside in New Zealand, the Polynesian colonists of AD 1250–1300 from the Society Islands, familiarised themselves with New Zealand rock resources. They quickly discovered obsidian at Mayor Island, near Taupo, and in the Kaeo–Kerikeri area of Northland. Soon afterwards, they discovered pounamu and argillite (pakohe) in the South Island. Apart from these highly prized rock types, they also exploited all others as it suited them.

The Second Wave from Europe

Then came the Europeans. The march of western philosophy and civilisation brought science, and the result today is a deep appreciation of ancient New Zealand. However, before wisdom was gained about just how finite natural resources are on a relatively small island, we humans systematically introduced a staggering diversity of European, Asian (tahr) and Australian animals, birds and plants. One of the more damaging has been the possum.

10. The upper Taieri River meanders through part of central Otago, the product of an ever changing landscape and climate. This is just one of the many remarkable landscapes that make up the modern face of the New Zealand topography. There is endless fascinating variety, and all within such a small country. This bewildering variety is due in part to our modern-day tectonic setting and in part to our geological heritage, in other words from what we have inherited through our long history.

11. Kiore, the Pacific rat, has arguably made the greatest impact in recent times on New Zealand's flora and fauna. Carried by Maori on their canoes as a food, kiore proliferated in a new land where there were few predators nor significant competition from other animals. A recent theory that kiore may have arrived 2000 years ago has now been discredited, with scientific consensus settling on around the year AD 1300 for both Maori and kiore arrival.

12. Pollen: the product of the male sex organ in flowering plants, the preserve of palynologists, the most abundant and widespread terrestrial organic material preserved in sediments and sedimentary rocks, and the source of much human misery by virtue of its allergen qualities. This exquisite specimen is pollen of the common dandelion. It is tiny, only 0.003 mm in size, and is typical of pollen of many flowering plants. It is not unusual for pollen counts to be as much as 30,000 grains per cubic centimetre of air. This remarkable stuff is tough, and yet easily extracted from rock. Evidence from pollen studies is transforming our knowledge of ancient New Zealand.

13

14

Road Kill Possum

To close this book on a lighter note, here is the story of the fossil possum on display at Te Papa – a modern fossil, no less.

While on a field trip in 1998 near Acacia Bay to the west of Taupo, one of us (HJC) spotted a partial skeleton preserved in the road – the actual bitumen road surface. The object of the field trip was to examine lake sediments that accumulated in Lake Taupo when it was 34 metres higher, immediately after the most recent Taupo eruption about AD 200. But the fossil possum had nothing whatsoever to do with this.

It was such a handsome specimen! The Taupo District Council was approached and the roading company Fulton Hogan was instructed 'to assist in the removal of an object from the road surface'. This is exactly what happened. A neat square of the bitumen was cut out and then the road was repaired. The bitumen was only 5 centimetres deep and is known as 'low-cost roading'.

How did the possum get there? You might think that road construction engineers leave puddles of tar for unsuspecting animals to get trapped in, but this was not the case. Possums are struck by vehicles and killed. Their carcasses remain on the roadside and they decompose. Resultant putrid fluids then interact with the tar and behave as a solvent, softening it. Vehicles continue to run over the dead possum and force the bones into the softened tar. In time, the flesh is totally decomposed but the bones are preserved within the bitumen road surface. In a way, dead possums behave rather like poultices, drawing tar to the surface of the road. At the time of discovery in 1998, the road surface was just six years old, so this particular possum was probably born about 1992.

It counts as one of the youngest fossils known from New Zealand, and exemplifies the preservation potential of tar. One of the most famous fossil localities in the world, the La Brea Tar Pits in Los Angeles, is in natural asphalt. This locality has produced abundant fossils of dire wolves and other scary animals of Late Pleistocene age.

Ancient New Zealand

And so ends our story, with a fossil possum no less. And yet it seems fitting, perhaps symbolic of a future pristine, rehabilitated New Zealand, free of pesky undesirable colonists and immigrants. How better to end a short discourse on key aspects of the history of New Zealand, and especially the 505 million years of history pertaining to New Zealand's geological heritage.

And surely the fossil possum may also be symbolic of how things sometimes really are: the result of a random sequence of accidents, chance events and lucky experiments. It is also fitting that the possum heralds from Australia. After all, we now know that New Zealand is largely comprised of greywacke, originally a great pile of muddy sands that is probably derived from Queensland. Furthermore, Zealandia rifted away from the Australian sector of Gondwanaland. In this sense, New Zealand is substantially Australian; Australia is where we are from, or at least in large part.

The past 120,000 years seems trivial compared with the much greater 505 million years of our known geological history, and yet in terms of human presence in New Zealand, just 700 years or so, 120,000 years is about 25 times less trivial. It begins to have meaning.

This highlights the most difficult aspect of geology and geological history that any of us experience: time. How do we conceptualise long spans of time? We have difficulty enough processing dozens of years in our minds let alone hundreds, thousands and millions. There is no easy solution, but geologists just get used to it, and therefore so might anyone.

The keys to our search for ancient New Zealand have included geological mapping; a deep geological knowledge of fluids, minerals, rocks, fossils and their formation; a thorough understanding of the orderly history of life; and access to modern isotopic dating techniques. Armed with these prerequisites, the memory banks that retain the long history of our land can be and are being read with

13. This cut piece of 'low-cost roading' from Acacia Bay, near Taupo, was laid down in 1992. Yet it sports the remains of a possum skeleton. On display at Te Papa, it is affectionately known as 'the fossil possum'. It may not rate as a true fossil of any great antiquity, but it serves to highlight the preservation qualities of tar. The bones and teeth are beautifully preserved, in much the same way and in similar medium, as the fossil sabre-toothed tigers and dire wolves in the La Brea Tar Pits in Los Angeles.

14. The 'petroglyphs' at Te Ana-a-Moe Cave, on the western shore of Te Whanga Lagoon, Chatham Island. These are interpreted as rock carvings crafted by Moriori, the indigenous people of the Chatham Islands who arrived about AD 1450–1500 (from mainland New Zealand), some 150–200 years after Maori arrived in New Zealand (about AD 1300). The carvings are in soft Eocene limestone.

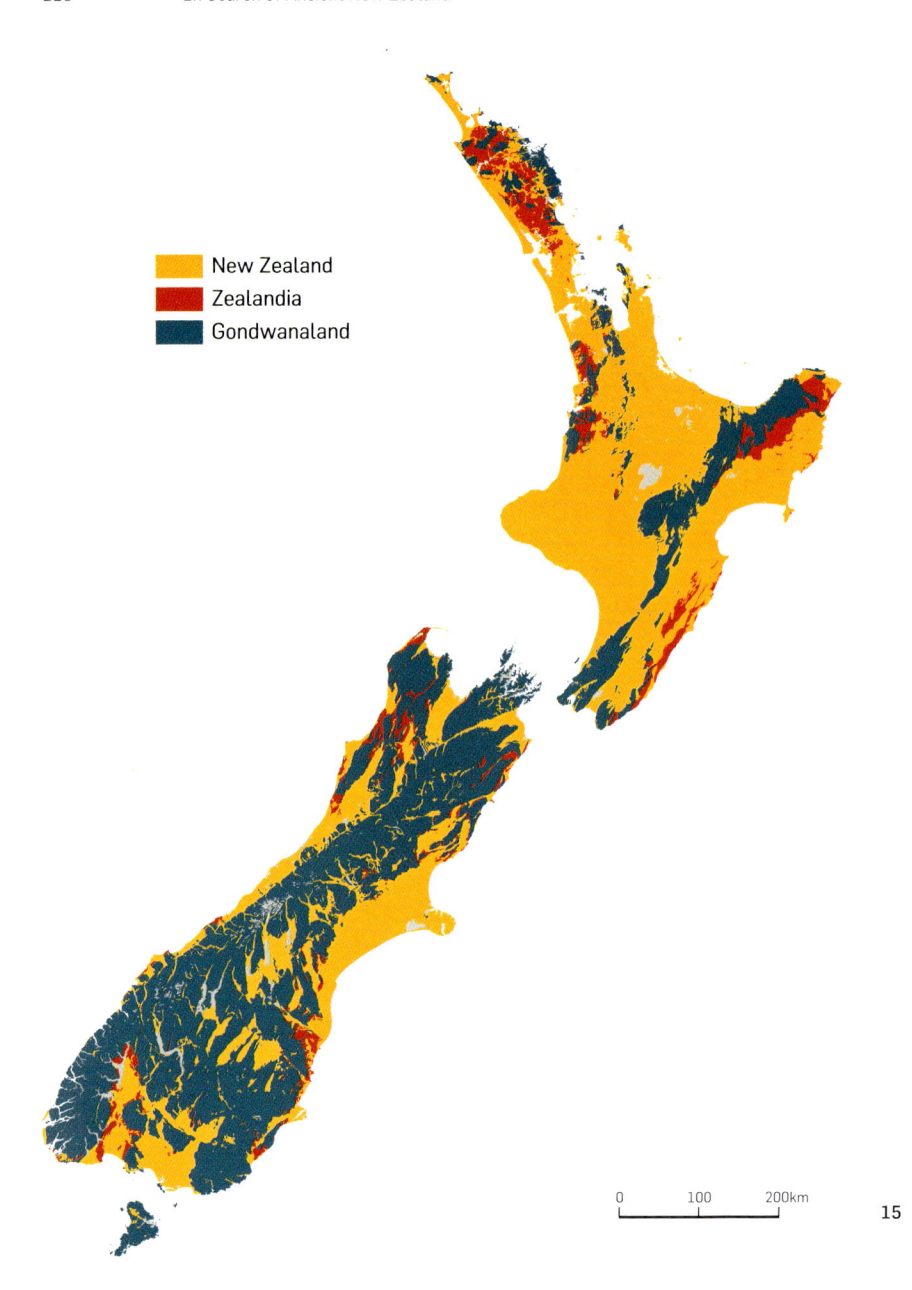

increasing ease. We have unravelled aspects of more than 500 million years of New Zealand's heritage, its origins, its whakapapa.

In summary, we can trace the geological origins of New Zealand back to more than 3 billion years ago, to the earliest continental crust, when the Earth was but an adolescent.

Our ancient history, representing much of Paleozoic and Mesozoic time, from 505 to 83 million years ago, is that of the eastern margin of Gondwanaland. Then, with the rifting of Zealandia 83 million years ago, our continent departed from the mother-ship of Gondwanaland in Late Cretaceous time. This journey of independence and increasing isolation began promisingly but Zealandia was stretched, thinned and slowly lost buoyancy. It sank, 2–3 kilometres over a period of about 60 million years. Zealandia diminished in land area to a low point 23 million years ago during earliest Miocene time. It is conceivable that it sank

15. A glimpse of New Zealand today showing the distribution of rocks and sediments that relate to the three major phases of our geological history: Gondwanaland (505-83 Ma; blue), Zealandia (83-23 Ma; red) and New Zealand (23-0 Ma; yellow).

completely, thereby erasing all terrestrial Zealandian life. Then, quite suddenly, the modern plate boundary developed through Zealandia and became a vigorous collision zone within the crust. This resulted in mountain building, the pushing up of continental crust, giving rise to New Zealand as we know it today.

We have entertained the possibility that New Zealand arose from a state of total immersion, total inundation by the sea. An obvious implication of this is that all New Zealand's native plants and animals have somehow arrived and evolved within the past 23 million years.

This book may be regarded as a popular account of New Zealand's geological history and yet it is different from any previous book on this topic in that it presents a more complicated history. First there was Gondwanaland, but then there was Zealandia. Real New Zealand only begins to appear about 23 million years ago. This new thinking reflects the ever-changing nature of science; it is never still.

A particular emphasis in this book has been a consideration of how old things are in our geological landscape, and how we know they are that old. But equally important are the questions: how did it form? What was the process and how did it get there?

Most significantly, for the first time in public, this book has raised the conceptual importance of Zealandia as a substantial fragment of continental crust, and the recognition of New Zealand as just the emergent fraction of a now largely sunken eighth continent. This new view of our part of the world will have profound implications for all of us. It is as if the New Zealand land mass has suddenly become something very different and much bigger than it was.

To better comprehend and understand the significance of this explanation, we have made good use of perspectives from the Chatham Islands, the most stable and most Australia-like part of New Zealand. We have also considered a number of extremely interesting and important new fossil localities that have been discovered within New Zealand during the past few decades.

Our quest, the search for ancient New Zealand, also serves to highlight the importance of fossils. Without them, we would know almost nothing about our ancient heritage. Modern paleontology is more exciting than ever before because so many new techniques and technological advances are available. Once the preserve of princes, officers and gentlemen, paleontology is now in the hands of some of the world's most accomplished scientists and most sophisticated and advanced research academies. The thirst for the excitement and new insight that fossils have to offer never ends.

This is the big picture, at least as good as we can get it so far, and with this understanding we can see the future, because to know the past is to know the way forward. To know history is to know the future.

The story of our long history relates ultimately to processes operating within the mantle. The surface geology of New Zealand has revealed to us the endless struggle of the crust in response to thermal processes associated initially with Gondwanaland, then Zealandia and now New Zealand. To use these terms is to invoke long-term events and effects operating within the mantle. And this realisation points the way forwards in our understanding of what will happen to New Zealand in the future. The answer lies in the mantle. Its behaviour is all-controlling. A new generation of research will lead us forwards so that in time, with sophisticated modelling and computing, we will be able to hazard a guess or even predict how New Zealand will change geologically and might look in the thousands and millions of years ahead.

One thing is for certain: things will change. Who knows, the plate boundary may step eastwards (again) leaving New Zealand to slowly sink back to the ambient base level of greater Zealandia, some 1000 to 2000 metres below the sea. Or perhaps it may step west, with a similar result. It is only by chance, a quirk of fate, that the plate boundary cut its way through Zealandia rather than around it and only by chance that New Zealand was pushed up. It is just an edge effect, but a pleasant one and a very interesting one, at least for the time being.

Picture Credits

Inside Cover: Bleached skeletons of modern bryozoans from New Zealand. These are colonial marine animals, distant relatives of corals that grow a skeleton comprised of calcite. They come in many shapes and patterns forming threads, nets, sticks, laces, films, bushes and clumps. They are perhaps best known as the plant-like organisms that foul wharves and the undersides of ships. They are common in the seas around New Zealand in water depths of less than 200 metres, and abound in water depths of about 80 metres. Fossil bryozoans are an immensely significant component of limestones of Cenozoic age in New Zealand. Oamaru Stone is largely comprised of bryozoan fossils.

Part One Opening Image: In 1864 Austrian geologist Ferdinand von Hochstetter named a peculiar rock he had discovered on Dun Mountain, near Nelson. He named it dunite, after the mountain. We now know that it represents ancient oceanic crust that forms a distinctive belt of rocks that can be traced throughout New Zealand.

Part Two Opening Image: *Allisporites australis* pollen of a Jurassic seed fern. This specimen is from a sample collected from Murihiku terrane rocks in Southland. Much of our knowledge of the plant life of the Gondwanaland heritage of New Zealand's history is based on palynology. There is an excellent Triassic, Jurassic and Cretaceous record of fossil spores and pollen. The oldest known fossil pollens and spores from New Zealand rocks are of Late Permian age. The photograph was taken down a microscope under strong magnification(x 4000).

Part Three Opening Image: Hawkdun Range, central Otago. The magnificent flat surface on the crest of the range is a well-preserved remnant of a much more extensive surface that was cut by the sea during Eocene time as Zealandia slowly sank. New Zealand as we know it has subsequently been pushed up by tectonism associated with plate boundary collision.

Part Four Opening Image: A superbly preserved fossil fish, found in muddy siltstone near Ormond, Gisborne. These lake sediments are of middle Pleistocene age, about 1 million years old. Found by Liz Kennedy (GNS Science) while sampling for fossil leaves and pollen, this is regarded by Bob McDowall (NIWA), an authority on New Zealand native freshwater fish, as the only known fossil grayling. One of the largest of New Zealand's native fish, the grayling became extinct in the 1920s. Length of fossil skeleton: 12 cm.

FRONT SECTION
Inside cover: Dennis Gordon, NIWA
pp. 6–7: 1/2-004109-F, Alexander Turnbull Library, Wellington
pp. 8–9: Lloyd Homer, GNS Science
p. 11: Margaret Low, GNS Science
pp. 14–15: Lloyd Homer, GNS Science
p. 18: TBC

PART ONE: INTRODUCTION
Lloyd Homer, GNS Science

CHAPTER ONE
1. PUBL-0023-001, Alexander Turnbull Library, Wellington
2. GNS Science

CHAPTER TWO
1. NASA/JPL–Caltech
2. Kate Whitley, GNS Science
3. GNS Science
4. John Rogers
5. Lloyd Homer, GNS Science
6. James Jackson
7. GNS Science
8. GNS Science
9. GNS Science
10. Lloyd Homer, GNS Science
11. GNS Science
12. GNS Science

CHAPTER THREE
1. Lloyd Homer, GNS Science
2. A & B: Nick Mortimer, GNS Science
3. GNS Science
4. Lloyd Homer, GNS Science
5. GNS Science
6. Graham Leonard, GNS Science
7. Lloyd Homer, GNS Science
8. GNS Science
9. Lloyd Homer, GNS Science
10. Hamish Campbell, GNS Science
11. Nick Mortimer, GNS Science
12. Lloyd Homer, GNS Science

PART TWO: GONDWANALAND
Ian Raine, GNS Science

CHAPTER FOUR
1. GNS Science
2. Lloyd Homer, GNS Science
3. Krzysztof Pfeiffer, Auckland Museum
4. Lloyd Homer, GNS Science
5. Andy Tulloch, GNS Science
6. PAColl-6238-32, Alexander Turnbull Library, Wellington
7. Hamish Campbell, GNS Science

CHAPTER FIVE
1. GNS Science
2. Lloyd Homer, GNS Science
3. Lloyd Homer, GNS Science
4. Lloyd Homer, GNS Science
5. GNS Science
6. GNS Science
7. John Simes, GNS Science

CHAPTER SIX
1. GNS Science
2. Heidi Schlumpf, GNS Science
3. Lloyd Homer, GNS Science
i. Malcolm Simpson, Private Collection
ii. Lloyd Homer, GNS Science
4. Robyn Cooper
5. Kate Whitley, GNS Science
6. Kate Whitley, GNS Science
7. Heidi Schlumpf, GNS Science
8. John Simes, GNS Science

CHAPTER SEVEN
1. GNS Science
2. Hamish Campbell, GNS Science
3. Kathryn Tulloch
4. GNS Science
5. Lloyd Homer, GNS Science
6. Lloyd Homer, GNS Science

CHAPTER EIGHT
1. John Simes, GNS Science
2. Lloyd Homer, GNS Science
3. Wendy St George, GNS Science
4. Catherine Waterhouse
5. Ewan Fordyce, University of Otago
6. Kate Whitley, GNS Science
7. Dallas Mildenhall, GNS Science
8. Lloyd Homer, GNS Science
9. Ewan Fordyce, University of Otago
10. Hamish Campbell, GNS Science
11. Rosemary Campbell
12. A & B: Ewan Fordyce, University of Otago
13. Lloyd Homer, GNS Science
14. Louise Cotterall, University of Auckland
15. Ewan Fordyce, University of Otago
16. Margaret Low, GNS Science
17. Vivi Vajda
18. Ewan Fordyce, University of Otago
19. Lloyd Homer, GNS Science
20. John Simes, GNS Science
21. Lloyd Homer, GNS Science
22. Susan Maclaurin, Massey University
23. John Simes, GNS Science
24. John Simes, GNS Science
25. John Callan, GNS Science
26. The *Dominion*
27. Brendan Hayes
28. Louise Cotterall, University of Auckland
29. Lloyd Homer, GNS Science
30. Tony Edwards, GNS Science
31. John Simes, GNS Science
32. John Simes, GNS Science

PART THREE: ZEALANDIA: 83–23 MILLION YEARS AGO
Lloyd Homer, GNS Science

CHAPTER NINE
1. GNS Science
2. Te Papa Tongarewa Museum of New Zealand
3. Te Papa Tongarewa Museum of New Zealand
4. Te Papa Tongarewa Museum of New Zealand
5. Jenny Worthy
i. Lloyd Homer, GNS Science
ii. Te Papa Tongarewa Museum of New Zealand

iii. Te Papa Tongarewa Museum
of New Zealand
6. Lloyd Homer, GNS Science
7. Paul Schiøler, GNS Science
8. NASA/JPL–Caltech
9. Hamish Campbell, GNS Science
10. Chris Consoli, Monash University
11. Margaret Low, GNS Science
12. Hamish Campbell, GNS Science
13. Don Davis/NASA

CHAPTER TEN
1. Lloyd Homer, GNS Science
2. GNS Science
3. Lloyd Homer, GNS Science
4. Lloyd Homer, GNS Science
5. Lloyd Homer, GNS Science
6. Lloyd Homer, GNS Science
7. NASA/JPL–Caltech
8. GNS Science
9. PA1-o-326-09,
Alexander Turnbull Library,
Wellington
10. GNS Science
11. Lloyd Homer, GNS Science
12. GNS Science

CHAPTER ELEVEN
1. Hamish Campbell, GNS Science
2. Hamish Campbell, GNS Science
3. Lloyd Homer, GNS Science
4. GNS Science
5. Liz Kennedy, GNS Science
6. Doug Campbell, University of Otago
7. A & B: Tony Harris, Otago Museum
8. Lloyd Homer, GNS Science
9. Tony Harris
10. John Simes, GNS Science
11. John Simes, GNS Science
12. *Waikato Times*
13. Ewan Fordyce, University of Otago
14. Hamish Campbell, GNS Science
15. Margaret Low, GNS Science
16. Lloyd Homer, GNS Science

CHAPTER TWELVE
1. Kate Whitley, GNS Science
2. 1/2-025368-F,
Alexander Turnbull Library,
Wellington
3. Lloyd Homer, GNS Science
4. GNS Science
5. Neil Campbell
6. Lloyd Homer, GNS Science
7. Lloyd Homer, GNS Science
8. John Simes, GNS Science
9. Lloyd Homer, GNS Science
10. A: Lloyd Homer, GNS Science
10. B: Dennis Gordon, NIWA
11. John Simes, GNS Science
12. Ewan Fordyce, University of Otago
13. Ewan Fordyce, University of Otago

PART THREE: NEW ZEALAND
Liz Kennedy, GNS Science

CHAPTER THIRTEEN
1. NASA/JPL–Caltech
2. GNS Science
3. Ewan Fordyce, University of Otago
4. Lloyd Homer, GNS Science

i. Hamish Campbell, GNS Science
5. Mike Aviss, Department of Conservation
6. S Lindsay & Sue Hand
7. Dick Veitch, Department
of Conservation
8. Daphne Lee, University of Otago
9. Ewan Fordyce, University of Otago
10. GNS Science
11. Lloyd Homer, GNS Science
12. Lloyd Homer, GNS Science
13. Lloyd Homer, GNS Science
14. GNS Science
15. Lloyd Homer, GNS Science
16. Jennifer Bannister, University of Otago
17. John Simes, GNS Science
18. John Simes, GNS Science

CHAPTER FOURTEEN
1. GNS Science
2. GNS Science
3. Neville Taylor, PandIS, Cambridge
University
4. GNS Science
5. GNS Science
6. Lloyd Homer, GNS Science
7. GNS Science
8. Hamish Campbell, GNS Science
9. Tui de Roy
10. Rob Suisted, www.naturespic.com
11. Hamish Campbell, GNS Science
12. Lloyd Homer, GNS Science
13. Hamish Campbell, GNS Science
14. John Simes, GNS Science
15. Heidi Schlumpf, GNS Science
16. Tony Eldon, Department
of Conservation
17. Paddy Ryan, Department
of Conservation
18. A & B: Puke Ariki Museum

CHAPTER FIFTEEN
1. Hamish Campbell, GNS Science
2. GNS Science
3. Lloyd Homer, GNS Science
4. Lloyd Homer, GNS Science
5. Jane Forsyth, GNS Science
6. John Simes, GNS Science
7. Rod Morris, Department of Conservation
8. Paul Shilov, Department of Conservation
9. Hamish Campbell, GNS Science
10. Lloyd Homer, GNS Science
11. Rod Morris, Department of Conservation
12. Ian Raine, GNS Science
13. Hamish Campbell, GNS Science
14. Hamish Campbell, GNS Science
15. GNS Science

Index

Abbotsford Formation 157, 159
Acrospirifer coxi (lampshell) 83
Adams, Chris 60, 63, 62, 70, 77
Africa 28, 30, 72
Aitcheson, Jonathan 135
Akaroa Volcano 149, 202–3
Albatross Point, Kawhia Harbour 104, 109
algae 74, 104, 112, 120, 122, 129, 155–6, 185
Allan, Robin 82
allochthons 186–7, 187
alpine areas 105, 208, 217–18
Alpine Fault 39, 68–9, 76, 133, 136, 138, 139–41, 139, 140, 142, 179, 203–5, 204
Altiplano, South America 32, 34, 50
aluminium 28, 30, 38, 40, 46
Alvarez, Walter 127
amber 150–1
ammonites 106–7, 107, 108, 109, 111, 112, 117, 126, 154
ammonoids 94, 99, 100, 103, 104, 106–7
amphibians 101, 107, 116
andesite 45, 54, 90, 113, 187
Anhanguera (pterosaur) 121
animals 16–17
    foreign 223–5
    fossils see fossils
    dinosaurs see dinosaurs
    Gondwanaland 96–7, 101–113
    New Zealand 180–6, 194, 210–12
    Zealandia 116–29, 149–53
    see also mammals; reptiles; specific animals
Anatoki Range, Tasman Mountains 59
Anatoki Thrust 80
Andrill Project 156
Antarctic ice sheet 198
Antarctic Plate 36
Antarctica 28, 30, 72, 78, 85, 153, 156
Aoraki (Mt Cook) 19, 105, 139, 206
aragonite 211
Archer, Brenda and Phil 60
Archer, Michael 185
Arctica 28, 30
Arrow Rocks, Whangaroa Bay 94, 98–9, 100, 101
Arthur's Pass 73, 105
ash 48–51, 49, 129, 144, 146, 189, 202, 219
Ashburton River 111
asteroids 27
astronomical surveying 22, 25
Atlantica 28, 30
atomodesmatinid clams 98
Atsushi Takemura 98, 100
Auckland Island 45, 91, 194
Auckland volcanoes 42–3, 42, 43, 189, 222–3
Australia 28, 30, 60, 72, 81–2, 85, 96, 101, 104, 225
Australian and Pacific plates 35, 36, 54, 137–8, 139, 140, 178–80, 186–7, 191, 203–5
    see also Alpine Fault; Southern Alps
Awakino Gorge 105
axial ranges 47, 64, 66–7, 136, 137, 179, 191, 197

bacteria 74, 151, 161, 185
Baltica 28, 30
Bam Earthquake, Iran 138, 139
Banks Peninsula 45, 149, 194, 202–3
Bannister, Jennifer 185, 186
Bannockburn, central Otago 152
barnacles 156, 168
Barnicoat Range, Nelson 97
Barretts Formation, Southland 111
Barrier Range, Fiordland 111
basalt 26–30, 38–46, 40
    flood basalt 46, 78
    large igneous provinces (LIP) 46, 98
basement rocks 47, 65, 68, 74–5, 76, 84, 92, 97, 99, 105, 112, 201
batholiths 46–7
    Median Batholith 46–7, 76, 77, 84–91, 85, 86, 87, 88, 89
Bathysiphon (tube-forming organism) 73
bats 182, 184
Bay of Plenty 47, 48
Beck, Russell 185
Beehive Island, Auckland 190
Begg, John 162
belemnites 108, 109, 112, 154
Benmore Dam, Otago 101
Benson, Noel 79, 81
Beu, Alan 209, 212
biota
    New Zealand 210–12
    Zealandia 116–19
birds 107, 111, 116–17, 151, 153, 180, 220–2
    bony-toothed pelican 213, 213
    penuinoid birds 155, 155, 168, 174–5
bivalves
    Gondwanaland 71, 77–8, 82, 94, 96, 97, 98, 103, 104, 108, 111, 112
    New Zealand 209, 212, 194
    Zealandia 126, 153, 154, 174
Black, Phillipa 61
Black Robin 221
black sand 189
Black Snipe 221
Blind Jim's Creek, Chatham Island 157
Bluff 86
'boundary clay' 127, 129
Bounty Islands 46, 47, 91, 207
brachiopods 11, 199
    Gondwanaland 77, 79, 82, 83, 94, 96, 101, 103, 104, 104, 106, 108, 111, 112
    Zealandia 147, 154, 168
Bradshaw, Margaret 82
Brazier, Ron 212
Brook Street terrane 92, 94, 97, 108, 111
Browns Island 42
Brunhes Event 200
bryozoans 77, 82, 84, 104, 111, 154, 156, 168, 185
Buller terrane 76, 77, 80, 82

calcite 96, 107, 108, 112, 163, 173, 183, 199
calcium 30, 38, 40, 75, 174, 198, 200
calderas 48–9, 51, 54, 55
Cambrian age 78, 79
Campbell, Doug 100, 105, 153
Campbell, Hamish 101, 153
Campbell Island 25, 45, 58, 77, 194
Campbell Plateau 24, 36, 37, 91, 97
Canterbury (Rakaia terrane) 92, 94, 97, 101, 103, 105
Canterbury Plains 68, 112–13, 138, 149, 218
Cape Foulwind 175
Cape Turakirae 134–5
Caples terrane 92, 94, 97, 99, 103
carbon dating 144, 214–5
Carboniferous age 92–4

Carcharodon megalodon (shark) 174
Caroline Bay, Timaru 45, 219
Cascade Valley, Fiordland 138
Castle Hill, Southland 173, 204
Castle Hill Basin, Canterbury 170–1, 173
Casuarina 152
Catlins coast 99, 102, 110, 111, 127
caves 171–3, 220–1
central Otago 120, 146, 148, 152, 165, 194, 222
    St Bathans, Lake Manuherikia 120, 180–5, 182
cephalopods 94, 108
Challinor, Brian 108
Chatham Island Petrel 221
Chatham Islands 24, 45, 49, 113, 123, 124, 125, 147, 207–10, 238
    Chatham Island 124, 126, 147, 148, 156, 157, 206, 209, 210
    Dieffenbach's Locality 156, 157
    Momoe-a-toa 209
    Ohira Bay 148
    Red Bluff Tuff Formation 147
    Te Ana-a-Moe Cave petroglyphs 220–1, 224
    Te Whanga Lagoon 156–7, 220–1, 225
    Tiorori 126, 154
    birds 221
    dinosaurs 123, 124–6, 124
    Pitt Island
        Flowerpot Harbour 154
        Hakepa Hill 148
        Moutapu Point 210
        Takatika Grit 123, 124–6
        Tupuangi Formation 113, 125–6, 125, 127
        Waihere Bay 113, 125, 125, 127
    Southeast Island 221–2, 221
Chatham Rise 24, 36, 37, 113, 125–6, 147, 210
Chilean Earthquake 139
China 61, 98
Chicxulub Meteorite, Mexico 98, 127–9, 128
Christoffel, David 30
clams 71, 96, 98, 103, 104, 112, 153
Clavigera bisulcata 11
Clent Hills Group 111
climate 112, 153, 197, 198–200, 203, 212, 215–17, 219, 220
    ice ages 97, 198, 204, 208, 216, 217–18, 219
    subtropical conditions 152, 169, 194, 195
coal 39–40, 93, 126, 150–1, 159, 171, 180, 191
Cobb Valley, Nelson 78, 78, 79, 81
cobbles 113, 124
Cocos zeylandica (coconuts) 195
coconuts 155, 194, 195
comet impact theory 98
concretion 157
conodonts 79, 80, 81, 82, 93, 94, 98, 99, 101, 103, 106
Consoli, Chris 124, 125
continent, definition 29
continental accretion 73
continental crust 22, 24, 26–37, 31, 32, 33, 71
    rifting see rifting/rift zones
continental drift 29, 32, 36
conulariids 94–6, 95, 104, 106
convection 36
Cook, Captain James 22–3, 23
Cook Strait 14–15, 103, 105, 106, 217
Cooper, Roger 80, 81

Index

Coopers Beach, Northland 195
coral 80, 82, 94, 111, 168, 169, 173, 194, 238
Coromandel Arc 191
Coromandel Peninsula 47
Coverham, north Canterbury 113
Crampton, James 112
Crawford, James Coutts 162
Cremnoceramus bicorrugatus (clam) 112
Cretaceous Period 112–13, 154
crinoids 103, 104, 111
crocodiles 180–2, 181, 184–5, 194, 221
Curio Bay, Catlins coast 110, 111

dating fossils, minerals, rock 30–1, 46, 60–3, 75, 77, 86, 91
  see also radiocarbon dating
de Jersey, Noel 104, 111
Desert Road 50
Devonian Period 82, 84
diatoms/diatomite 155–6, 185
Dicroidium flora 101, 103
Dieffenbach, Ernst 156
Dieffenbach's Locality, Chatham Island 156, 157
Dilophus campbelli (mayfly) 152, 153
Dimitobelus lindsayi (belemnite) 109
dinoflagellates 104, 111, 112, 120, 122, 123, 124, 154, 155
dinosaurs 97, 110–11, 116–29, 119
  Chatham Islands 123, 124–6, 124
  Gondwanaland 97, 108–9, 124–6
  Hawke's Bay 119–24
  Port Waikato 111, 124, 126–7
  Zealandia 116–29
Douglas, Barry 120, 182, 184–5
drill-core samples 43, 47–8, 149, 197, 199, 200, 203, 220, 222
Dun Mountain 20–21, 39, 92, 94
Dun Mountain-Maitai terrane 92, 94, 97, 99, 101, 140
Dunedin area 45, 154, 190
Dunedin Volcano 45, 148, 190, 194
dunite 40
D'Urville Island 94, 99

Earth 26–37, 32, 33
  creation stories 19, 58, 60
  crust 26–37, 38–40, 141–2, 205
  magnetic polarity 31–2, 36, 200–1, 200, 202
  mantle 32–6, 32, 33, 41, 141–2, 167, 227
  nuclear energy 36
  orbit 198, 199
earthquakes 42, 55, 130–45, 142
  Bam Earthquake, Iran 138, 139
  Chilean Earthquake 139
  Edgecumbe Earthquake 55, 132, 133, 143, 207
  Hauwhenua Earthquake 136
  Hawke's Bay Earthquake 137
  Kobe Earthquake, Japan 139
  Murchison Earthquake 132
  Richter Scale 139
  seismology 142–4
  shallow versus deep 142
  speed 142
  Sumatra Earthquake 139
  Wairarapa Earthquake 132, 135, 136, 141, 205
  see also faults
East Coast Allochthon 187

East Coast Basin 197, 202–3
Eastern province, Gondwanaland 75, 76, 77–8, 84, 90, 91, 92–113
  terranes 92–113
echinoderms 94, 147, 154, 155, 156, 168, 169, 173
echinoids 103, 104, 108
Edbrooke, Steve 162
Edgecumbe Earthquake 55, 132, 133, 143, 207
Edgecumbe Fault 132–3, 133, 143
eels 81, 82
Einstein, Albert 61–3, 62
Ekatahuna 111
elasmosaur (marine reptile) 118, 120, 121, 154
elements 38
Eocene age 155, 156–9
erosion 147, 148, 157–9, 163, 163, 165, 170–1, 197
Exclusive Economic Zone (EEZ) 17
extinctions (mass) 105–6, 127–9

Falconer, Robin 30
Farewell Spit 150
faults 130–45, 131, 145
  Alpine Fault see Alpine Fault
  blind 138
  Edgecumbe Fault 132–3, 133, 143
  Hope Fault 34–5
  Kelly Fault 140
  measuring activity 144
  normal 132–3
  Paeroa Fault 132, 133
  Poukawa Fault 137–8
  reverse 133–4
  Taupo Fault Belt 132
  transpressional 131, 133–9, 204–5
  Wairarapa Fault 131, 132, 133–6, 134–5, 141, 205, 222
  Wellington faults 131, 131, 136–7, 145
feldspar 30, 40, 46, 63, 68, 90
Fiordland 47, 80, 81, 84, 85, 88–9, 90, 111, 138, 153, 206
fish 81, 82, 94, 104, 111, 117, 120, 154, 155, 168, 174, 180, 185
  Banded Kokopu 21
  flat-fish 211, 212–13
  grayling 213
  hagfish (blind eels) 81, 93
  scales 94, 111
  whitebait 212
Fleming, Sir Charles 153, 154, 164, 165
flora and fauna see animals; plants; specific creatures
flowers 116, 151, 151, 223
folding 205
foraminifera (plankton) 72, 73, 94, 183, 199
Fordyce, Ewan 174–5, 175, 183
forest remains 111, 215, 216–17
  see also trees
fossils 161
  ammonites 106–7, 107, 108, 109, 111, 112, 117, 126, 154
  ammonoids 94, 99, 100, 103, 104, 106–7
  amphibians 101, 101, 151
  birds 120, 121, 173, 175, 182, 213, 213, 220–1
  coal 150–1
  coconuts 155, 194, 195
  conodonts (teeth) 79, 80, 81, 82, 93, 94, 98, 99, 101, 103, 106
  corals 173

crustaceans 104, 108, 109, 154, 155
dinosaurs see dinosaurs
fish 211, 212–13
flowers 151, 151, 185
fruit 185
graptolites 81–2, 83
heart urchins 168, 173
insects 110, 111, 152, 153, 185
leaves 96, 97, 111, 151, 185, 194
lobster 109
'macrofossils' 79, 151, 183, 189
mammals 120, 151, 153, 182–4, 184
microfossils see microfossils
moa 182, 220
paua 211
penguins 155, 155, 168, 174–5
plants 96–7, 96, 101, 103, 194
pollen and spores see pollen; spores
recycling/reworking 97, 124
reptiles (marine) 101, 104, 104, 111, 112, 121, 126, 154
  crocodiles 180–2, 181, 184–5, 194
  elasmosaurs 118, 154
  mosasaurs 119
  plesiosaur 121
reptiles (flying) 121
scallops 174, 209, 212
shark teeth 147, 154, 156–7, 157, 173, 174, 175
shellfish see molluscs; brachiopods
shellbeds 77–8, 82, 83, 94, 96, 101, 103–5, 108, 209, 212
soils 111, 202, 215, 216, 219
spores 97, 104, 106, 112, 149–50, 151, 153, 161, 185, 194
trace fossils 154
trees 110, 111, 125, 215, 216–7
trilobites (crustaceans) 77, 78
turtles 121, 153, 155, 168, 174
vertebrates 71, 111, 121, 122, 168, 184–5
whales 155, 168, 174–5, 183, 184
wood 38, 97, 111, 113, 151, 153, 161
fossil extraction 161
fossil hunting 183
Fossil Record File 105

gabbro 86, 90
gas 38–41, 49, 50, 54, 155, 159
gastropods (snails)
  Gondwanaland 77–8, 82, 94, 103–4, 111, 112
  New Zealand 194
  Zealandia 126, 154, 156, 173, 195
geographic positioning systems (GPS) 54
geological mapping 161–2
Geomagnetic Polarity Timescale 200
glaciation see ice ages
glacial landscape 217–18, 217
glauconite 124, 126, 165
gneiss 90
GNS GeoNet 42, 54, 139
GNS Science 16, 54, 77, 85, 91, 98, 139, 150, 161, 162, 206, 212
Glossopteris flora 96–7, 96
Gog Magog 87
Gondwanaland 37, 69, 71–3, 72
  animals, plants 96–7, 101–13
  Cambrian age 78, 79
  Carboniferous Period 92–4
  continental crust 71
  Cretaceous Period 112–13, 154
  Devonian Period 82, 84

dinosaurs 97, 108–9, 124–6
Eastern province 75, 76, 77–8, 84, 90, 91, 92–113
terranes 92–113
Jurassic Period 107–12, 126
marine life 78–9, 81–3, 93–113
Median Batholith 46–7, 76, 77, 84–91, 85, 87, 88, 89
Ordovician Period 78, 80–2
Permian Period 94–7
Silurian Period 82
Triassic age 73, 97–107
Western Province 74–91, 76
terranes 75–8, 76, 80, 82
Graham, Ian 77
granite 26–30, 38–9, 40, 46–8, 207
orbicular granite 90–1
plutons and batholiths 46–7, 86, 90
Grant-Mackie, Jack 103, 108
graptolites 81–2, 83
grass 116, 151, 217
Gray, John 156
Great Barrier Island 47, 191
Greenland Group, Westland 71, 80
greensands 125, 165, 180
greywacke 47, 64–73, 65, 68–9, 91, 204
Gulf of Thailand 36
Gyles, Jean 108

Hailes Quartzite formation, Takaka Valley 82
Hakepa Hill, Chatham Islands 148
Halcombe Hill 205
Haliotis (Notohaliotis) (paua) 211
Hand, Suzanne 185
Haremare Creek, Franz Josef 140
Harper Range, Canterbury 101
Harris, Tony 152, 153, 154
Haumuri Bluff, north Canterbury 108
Hauwhenua Earthquake 136
Hawaii 41, 42
Hawke's Bay 46, 153, 154, 183, 197, 211, 219, 220
dinosaurs 119–22, 153
earthquake 137–8
Mangahouanga Stream 118, 121, 183
Hayes, Brendan 108, 111, 126
Hector, Sir James 9, 162, 162
Herangi Range 153
Hikurangi LIP 46
Himalayas 204
Hokitika 140
Hokonui Hills, Southland 104, 105, 107
Hollyford Valley 88–9
Homer Tunnel, Fiordland 88–9, 90
Hope Fault 34–5
hot spots 19, 41
Houghton, Bruce 49
Hudson, Neville 108
Hunua Range 47
Hutt Valley 136–7, 205
Hutton, Frederick Wollaston 162
hydrocarbons 149–50, 159
hydrosphere 29

ice ages 97, 198, 204, 208, 216, 217–18, 219, 222
ichnofossils 154, 154
ichthyosaurs (marine reptiles) 104–5, 104, 119
Ida Valley 182
igneous rocks 30, 31–2, 34, 39, 84–5
large igneous provinces (LIPs) 46, 98

see also basalt; granite
ignimbrite 50–2, 228–9
Ikawhenua Range 111
India 28, 30, 46, 72
insects 108, 110, 151, 152, 153, 185, 220
iron 32–3, 38, 63, 189, 202
Isaac, Mike 162

Jackson Valley, Fiordland 138
Jenkins, Hugh and Graeme 94
John Inglis Valley, Fiordland 206
Jones, Craig 185
Jurassic Period 107–12, 126

Kaikohe 45
Kaikoura Mountains 34–5, 112
Kaimanawa Range 66–7, 68
Kaimanawa Wall 51, 54
Kaipara Volcano 189
Kaiwara, north Canterbury 111
Kaka Point, Catlins coast 99, 99, 101–3, 102
Kakahu 94
Kapiti Island 104, 221
Karamea 86, 90
Karamea Batholith 46
Karamea Granite 90, 91
karst/sinkhole topography 163
Kauri gum 150–1
Kawhia Harbour 104, 107, 108, 175
Kawhia Syncline 104
Kelly Fault 140
Kennedy, Liz 151, 213
King Country coast 103
kiore (rats) 184, 221, 223, 223
Kiritehere 103
kiwi 108, 220
Kobe Earthquake, Japan 139
Krakatoa, Indonesia 55

lahars 187–9
lake sediments 120, 153, 180–6, 182, 187, 213, 225
Lake Aviemore 94
Lake Manuherikia 180–2, 182
Lake Tarawera 44
Lake Taupo 47, 48, 48
Lake Te Anau caves 171–3
Lake Tutira 220
Lake Waitaki 195
Landis, Chuck 165, 165
landslides 206
large igneous provinces (LIPs) 46, 98
lava 45
leaves 96, 97, 111, 151, 185, 194
Lee, Daphne 185, 186, 209
Lee, Julie 162
Leonard, Graeme 43
Little Ben Sandstone, Nelson 99
limestone 39, 77, 79, 163, 168–9, 204
caves 171–3, 225
Eocene age 156–7, 224
Late Oligocene–Early Miocene formations 168–75, 168–9, 170–1
Pahau Late Jurassic 111–2
Lindqvist, Jon 120, 182, 184, 186
Livingstone 153
lizards see reptiles
lobster 109
loess 219–20, 219
Louisville Ridge 41–2
Louisville seamount chain 41
Lyell, Sir Charles 141, 141

Lyttelton 45, 149
Lyttelton Volcano 148, 194
Lytoceras taharoaense (ammonite) 107, 109

McDowall, Bob 213
MacFarlane, Donald 108
McKenzie, Dan 31
Mackenzie Basin 220
McNamara, James 185
maar 185
macrofossils 79, 151, 183, 189
see also fossils
magma 32, 43–55, 84–5, 149
magnesium 28, 30, 36, 38, 39, 40
magnetism of Earth 31–2, 36, 200–1, 200, 202
magnetostratigraphy 200–2
Mahinepua Peninsula, Whangaroa Bay 101
Maitai Group rocks 99, 140
Maitai River 94
mammals 97, 116, 120, 129, 151, 153, 182, 184, 184, 220
Mangahouanga Stream, Hawke's Bay 118, 121, 183
Mangakahu Valley 228–229
Mangakino Caldera 51
Mangere Island 194
Mangere Volcano, Chatham Islands 125, 194, 210
Mantell, Gideon 6, 141
Mantell, Walter 6, 141
Manukau Volcano 189
mantle 32–6, 32, 33, 41, 141–2, 167, 227
marble (Takaka) 80, 81
Marble Bay 94
Mariana Trench, Philippines 26, 32
marine fossils
Gondwanaland 92–113
New Zealand 197, 202–3, 212–13
Zealandia see Zealandia
see also fossils; microfossils; specific organisms
Marlborough 34–5, 47, 64, 68, 71, 112, 154
Maroa Caldera 54
Marokopa–Kiritehere coast 104, 105
Marshall, Patrick 50, 90
Maskelyne, Neville 25
Matai 215
Mataura Island 101
Mataura River 100
Mathias River 105
Matthews, Drum 31, 31
Matsuyama Event 200
Maui 19
Maungataniwha Formation 122
Maxwell, Phillip 111, 156, 156, 209, 212
mayfly 152, 153
Median Batholith 46–7, 76, 77, 84–91, 85, 86, 87, 88, 89
metamorphic rocks 16, 39, 63, 77, 84–5, 90
see also greywacke
metamorphism 75, 77
meteorites 60, 60, 98, 127, 128
mica 30, 46, 68, 70, 90
microfossils
Gondwanaland 71, 97, 98–9, 101, 103, 104, 105, 112
Zealandia 124, 151, 154, 161, 168
New Zealand 183, 189
see also conodonts; plankton; pollen; spores
Mid-Atlantic Ridge 31
mid-ocean ridges 31–2, 31, 32, 40–1, 71
Middlemarch Maar 185–6, 187

Index

Milankovitch, Milutin 198
Milburn 175
Mildenhall, Dallas 96, 96
minerals 38, 40
    see also specific minerals and rocks
mineral analysis 30–2, 60–3, 161
Miocene age 153, 164, 168–75
moa 108, 167, 180, 220
Moanasaurus mangahouangae (elasmosaur) 118
Moeraki Boulders 157–9, 158
Mohakatino Arc 189
molluscs
    Gondwanaland 71, 77–8, 79, 82, 94, 96, 97, 98, 103, 104, 108, 111, 112, 112, 117
    New Zealand 182, 194, 199, 209, 212
    Zealandia 126, 147, 153, 154, 156, 156, 168, 173, 174, 195
Molnar, Ralph 118, 185
Momoe-a-toa, Chatham Island 209
Monotis (clam) 103
Moon 26–8, 27
Moriori 225
Mortimer, Nick 85, 85
mosasaurs (marine reptiles) 119–20, 119, 121, 126, 129
Moturoa Island, Bay of Islands 94
Motutapu Island 223
Moutapu Point, Chatham Islands 210
Mt Aspiring National Park 39, 206
Mt Cook (Aoraki) 19, 105, 139, 206
Mt Edgecumbe (Putauaki) 19, 44
Mt Egmont (Taranaki) 19, 223
Mt Harper 101
Mt Horrible 45, 219
Mt Hutt Range 65
Mt Misery 166–7
Mt Pinatubo eruption 98
Mt Potts, Canterbury 101, 104–5
Mt St Helen's eruption 55
Mt Somers 112–13
Mt Tutoko 88–9, 90
mudstone 39, 82, 126, 153, 157–9, 164–5, 204
Murchison earthquake 132
Murihiku terrane 90, 92, 94, 97, 101, 103, 104, 105, 105, 107, 108
Musser, Anne 185

nanofossils 108, 112, 168
nautiloids 94, 104, 117, 126
Navy 30
Nelson area 59, 80, 82, 84, 94, 96, 97, 98, 99, 103, 109, 111, 166–7, 173
    Cobb Valley 78, 78–9, 81
    Golden Bay 77–8, 94, 98
Nena 30
New Caledonia 25, 85, 96, 155, 168
New Zealand 24, 75, 76, 226
    allochthons 186–7, 187
    animals, birds 165, 180–6, 194, 202–3, 210–12, 213, 220–2
    emergence 178–80, 181, 226–7
    last glacial period 215–8
    marine record 197, 212–3
    marine sediment 202–3
    Miocene age 164, 178–95
    plants 194, 210–12, 220–2
    Pliocene and Pleistocene ages 164, 196–7
    rifting see Taupo Volcanic Zone
    sediment distribution 181, 226
    tectonic activity 179–80, 222
    volcanic activity 187–94, 201, 207–10, 222–3
    within last 120,000 years 214–27
New Zealand continent see Zealandia
Ngai Tahu 19
Ngatoroirangi 19
Ngauruhoe 45, 54–5
Norfolk Basin, Northland 153
Nori Suzuki 100
North Island 47–8, 131, 137–8, 154
    compared to South Island 191
    volcanic activity 48–55, 187–91, 222–3
    see also Taupo Volcanic Zone
North Pole 31–2, 74, 200
Northland 45, 186–7, 187, 191, 194, 195, 197, 212, 223
    see also Waipapa terrane
Northland allochthon 187
Northland Volcanic Arc 187–9
Notohaglia 110
Nugget Point, Catlins coast 101, 102

Oamaru diatomite 155–6
Oamaru Stone (Ototara Limestone) 172, 173–4
obduction 186
obsidian 223
Ocean Beach, Wairarapa 215, 216
Ocean Drilling Programme 199
ocean floor see sea floor
oceanic crust 26–37, 31, 32, 33, 40–1, 71, 73, 113,
Ohira Bay, Chatham Islands 148
oil 39–40, 149, 155, 159
Okataina Caldera 49, 55
Oligocene age 153, 160–75
Oparara River, West Coast 90
Orakei Basin 43, 222
Ordovican Period 78, 80–2
Orongorongo River 103, 105
Oruanui Eruption 48–9
ostracods 104, 111
Otago (Rakaia terrane) 92, 94, 97, 101, 103, 105
Otago Harbour 45, 190
Otago Peninsula 190
Otamita Stream 104
Otamita Valley, Southland 105
Otematata, Canterbury 101
Owen, Stuart 99
oxygen 38–9, 98, 175, 185, 198–200
oysters 156

Pacific and Australian plates 35, 36, 54, 137–8, 139, 140, 178–80, 186–7, 191, 203–5
    see also Alpine Fault; Southern Alps
'Pacific Ring of Fire' 113, 187–9, 191, 207
Paeroa Fault 132, 133
Pahau terrane 92, 108, 111–12
Pakawau 151
Pakihi Island, Hauraki Gulf 106
Paleocene age 154–5
paleogeographic maps 164
paleontology 81, 99, 183, 227
    see also fossils
palynomorphs see pollen; spores
Panthalassa Ocean 73, 111, 112, 113
Papatuanuku (Earth Mother) 19, 58
Papua New Guinea 85
Paraconularia (conulariid) 95
Parapara Peak, Golden Bay 77–8, 94, 98
Park, James 195
paua 211, 212
peneplains 148, 165
penguins 155, 155, 168, 174–5
Pericosmos crawfordi (heart urchin) 168
peridotite 36, 40, 41
peripatus 108, 220
Permian Period 94–7
petrographic analysis 40
'petroglyphs' 224
Pihanga 19
pink and white terraces, Tarawera 45
Pitt Island see Chatham Islands
plankton 98, 100, 101, 120, 124, 149, 155–6, 168, 183, 198–200, 202
    algae and diatoms 74, 104, 112, 120, 122, 129, 155–6, 185
    foraminifera 72, 73, 94, 183, 199
    radiolarians 94, 98–9, 101, 103, 104, 105, 106, 106, 111, 154
plants 16, 17
    alpine 208
    Casuarina 152
    Dicroidium flora 101, 103
    ferns 116, 151
    flowers 118, 151, 151
    fruit 185
    Glossopteris flora 96–7, 96
    Gondwanaland 96–7, 101–113
    leaves 96, 97, 111, 151, 185, 194
    New Zealand 194, 210–12, 220–2
    pollen and spores see pollen; spores
    Zealandia 116–29, 149–53
    see also trees
plates 36
    Australian and Pacific see Australian and Pacific plates
plate tectonic theory 16, 30, 32, 34, 36, 42, 161, 165, 202–3
Pleistocene age 164, 196–7
plesiosaur 121, 175
Pleuronectiformes (flat-fish) 211, 212–13
Pliocene age 164, 196–7
plutonic rocks 46–7, 84, 86
plutons 46–7, 86, 90
polarity of Earth see magnetism
Pole, Mike 180, 184–5
pollen 223
    Gondwanaland 78, 97, 97, 103, 104, 105, 106, 111, 112
    New Zealand 183, 185, 194, 203, 210–12, 216, 220
    Zealandia 120, 149–50, 151, 153, 159, 161
Port Waikato 108, 111, 123, 124, 126–7
possums 223–5, 224
potassium 31, 36, 38, 46, 60–3, 86, 124, 126
Poukawa Fault 137–8
pounamu 77, 223
Prehistoric New Zealand (G Stevens et al) 16
Principles of Geology (C Lyell) 141
Productus Creek 94, 97
Pseudodontornis (bony-toothed pelican) 213
pterosaurs (flying reptiles) 121
Punakaiki (Pancake Rocks) 168–9, 169–71
Purerua Peninsula, Bay of Islands 94
Putauaki (Mt Edgecumbe) 19
Pycnodonte (Notostrea) *tarda* (oyster) 156

quartz 38, 46, 68, 69–70, 90, 125
Queensland 70

radioactive elements 30–1, 36, 48, 60–3, 75, 77
radiocarbon dating 144, 214–5, 221

radiolarians 94, 98–9, 101, 103, 104, 105, 106, 106, 111, 154
Raine, Ian 104, 104, 111, 153
Rakaia River 65, 105, 220
Rakaia terrane (Otago, Canterbury) 92, 94, 97, 101, 103, 105
Rakaihautu 19
Rakopi Formation, Pakawau 151
Ranginui (Sky Father) 19, 58
Rangitata Orogeny 107
Rangitoto Island 42, 223
Rastelligera elongata (brachiopod) 104
rats 184, 221, 223, 223
Rattenbury, Mark 162
Raukumara Peninsula 112, 186
Red Bluff Tuff Formation, Chatham Islands 147
Red Hills, Otago 39
Red Rocks Point 14–15
reefs 194
Reefton area 46, 71, 78, 80, 82
reptiles 107, 108, 116–22, 129, 154
 flying, terrestrial 116–17, 129, 184
  pterosaurs 121
 marine 101, 104, 111, 118
  elasmosaurs 118
  ichthyosaurs 104–5, 104, 119
  mosasaurs 119
  plesiosaurs 121, 175
  pliosaurs 120
Retallack, Greg 101
Retroceramus (clam) 71
rhyolite 45, 46, 48–51, 54, 55, 113, 207, 223
Rie Hori 98, 101
rifting/rift zones
 in North Island 36, 46, 48, 54, 133, 187, 207
 Zealandia from Gondwanaland 37, 47, 84, 86, 112, 132, 167, 180, 226
Rimutaka Range 47, 136
Roaring Bay, Nugget Point 102, 103, 104
Roberts, Edward 141
rocks 38–40
 analysis 60–3, 161
 distribution 150, 161–5, 181, 226
 igneous see igneous rocks
 metamorphic see metamorphic rocks
 sedimentary see sedimentary rocks
rock carvings 224
Rodinia 28, 30
Rogers, John 28
rostroconchs 94, 96
Rotorua 47, 55
Roxburgh 182
Ruahine Range 47, 136, 191
Ruapehu 48, 52–3, 54, 191–2, 223
Rutherford, Ernest 31, 60, 61

St Bathans, central Otago 120, 180–5, 182
sand dunes 218–19, 218
sandstone 39, 64, 82, 98, 103, 107, 119, 126, 164–5, 190, 192–3, 204
satellites 54
Satoshi Yamakita 100
scallops 174, 209, 212
scaphopods 94, 104
schist 47, 64, 204
sea floor 26, 147
sea-floor spreading 29, 29, 31–2, 36, 42, 73, 117, 191, 205
sea levels 32, 37, 167, 197–8, 199, 203, 215–17, 216, 219, 220
sea water chemistry 125, 161, 174–5, 198–200, 208

sediment 26, 68–73, 161, 197
sedimentary rocks 16, 39, 64, 68, 84–5, 90, 92, 97–9, 183, 202
 fossils in see fossils
 limestone see limestone
 mapping of distribution 150, 161–5
 mudstone 39, 82, 126, 153, 157–9, 164–5, 204
 sandstone see sandstone
 schist 47, 64, 204
 siltstone 39, 70, 103, 156, 164, 190, 192–3
 in Zealandia 160–75
 see also greywacke
seismology 142–4
Shackleton, Nick 200, 200
Shag Point, Otago 154, 175
sharks 82, 147, 154, 156–7, 157, 173, 174, 175
shellbeds 77–8, 82, 83, 94, 96, 101, 103–5, 108, 209, 212
shellfish see brachiopods, molluscs
Shore Plover 221
Siberian Traps eruption 46, 98
silica 45, 98, 111, 155–6, 198–200
siltstone 39, 70, 103, 156, 164, 190, 192–3
Silurian Period 82
Simes, John 81
Simplicites (ammonoid) 100
Simpson, Malcolm 79
sinkhole/karst topography 163
Snares (The) 207
sodium 30, 38, 40
Soddy, Frederick 31, 60, 61
soft-bodied animals 74
soil 111, 125, 141, 144, 148, 202–3, 215, 216, 219
Solar System 27, 60
Solvay Conference, Belgium 62
South Island 64–73
 compared to North Island 191
 Median Batholith 46–7, 76, 77, 84–91, 85, 86, 87, 88, 89
South Pole 31–2, 74, 94, 106, 111, 200
Southeast Island, Chathams Islands 221–2, 221
Southern Alps (Ka Tiritiri o te Moana) 19, 35, 68–9, 202–3, 203–5, 204
 see also Alpine Fault
Southland 95, 101, 103–5, 112, 126
 Catlins coast 99, 102, 110, 111, 127
Southland Syncline 90, 104, 107
Sphenodon punctatus (tuatara) 184
Spinomelon parki (gastropod) 195
sponges 74, 79, 80, 80–1, 82, 94, 126, 147, 154, 185, 219
spores
 Gondwanaland 78, 97, 103, 104, 105, 106, 111, 112
 New Zealand 185, 194
 Zealandia 149–51, 153, 161
Spörli, Bernard 61, 99
Squalodon (whale) 175
Stephens Island 101
Stephenson Island 105
Stevens, Graeme 108, 109
Stewart Island 47, 84–6, 87, 167, 207
Stilwell, Jeffrey 124, 125
stratigraphy 49, 203
 see also drill-core samples
Striatolamia macrota (sand shark) 157
strontium 75, 77, 174, 200
subduction 35, 45, 54, 73, 113, 178–9, 186, 191, 223
subduction zones 32, 191

subtropical conditions 152, 169, 194, 195
Sumatra Earthquake 139
Sun 26, 27, 58, 60, 61–2, 198–9, 199

Tahiti 22
Taiere River 222
Takaka marble 80, 81
Takaka terrane 76, 77, 78, 80, 82
Takaka Valley 82
Takatika Grit, Chatham Islands 123, 124–5
Takitimu Mountains, Southland 94, 95, 105
Tambora, Indonesia 55
Tane 19
taniwha 19
Taranaki 169, 213
Taranaki coast 105
Taranaki (Mt Egmont) 19
Tararua Range 47, 136, 218
Tarawera 44, 45, 55, 223
Tarawera Eruption 55
Taringatura Hills, Southland 16, 105, 173
Tasman Glacier 206, 217, 217
Tasman Formation 79
Tasman Mountains (Anatoki Range) 59
Tasman Sea floor 37, 41, 112, 167
Tasmania 46, 70, 77–8, 105
Tauhara 19, 113
Taupo Caldera 48, 54
Taupo Eruption 50
Taupo Fault Belt 132
Taupo Volcanic Zone 47, 48–50, 133, 143, 187, 189, 191–2, 207
Taupo–Napier Road 50, 212
Taylors Stream 105
Te-Ana-a-Moe Cave petroglyphs 220–1, 224
Te Anau 90
Te Hoata 19
Te Hoe Valley, Hawke's Bay 119–21
Te Kaukau Point 154
Te Kopia geothermal field 132
Te Kuiti 50–1
Te Pohue, Hawke's Bay 211
Te Pupu 19
Te Whanga Lagoon, Chatham Island 156, 157, 225
tectonic forces 130–2
tectonic plates 36
 Australian and Pacific see Australian and Pacific plates
 plate tectonic theory 16, 30, 32, 34, 36, 42, 161, 165, 202–3
tectonic uplift 75, 86, 159, 173, 180, 186, 189–91, 197, 203, 210, 216, 222
 see also Southern Alps
teeth 93, 104, 147, 154, 156–7, 157, 173, 174, 180
 conodonts 79, 80, 81, 82, 93, 94, 98, 99, 101, 103, 106
Tennyson, Alan 185
tephra 49, 51, 219, 222, 223
Terawhiti 68
terranes, Gondwanaland
 Eastern Province 92–113
 Western Province 75–8, 76, 80, 82
thermal processes 36, 37, 167, 198, 210, 227
 see also volcanoes
theropods 108, 111, 119, 121, 122, 123, 124, 126
Thorn, Vanessa 111
Three Kings Islands 60, 60
Tibetan Plateau 32, 34
Timaru area 45, 163, 219, 219
Timaru Basalt 219

Titahi Bay 215, 216
titanium 27, 189
Tonga–Kermadec Arc 187, 207
Tonga–Kermadec Trench 26, 40
Tongariro 19, 45, 54–5, 191–2, 223
Torlesse Range, Canterbury 73
Torlessia (tube fossils) 72, 73, 105
totara 215
Townsend, Dougal 162
trace fossils 103, 104, 153, 154–5, 154, 189
Tramway Formation, Nelson 99
trees 97, 110, 111, 151, 185, 194, 215
trenches 26, 32, 32, 71
Triassic Period 73, 97–107
trigoniids 71, 105,
trilobites (slaters) 77, 78, 79, 80, 94
Trümpy, Rudolf 96–7
tuatara 107, 167, 184, 220
Tulloch, Andy 85, 85
Tupaenuku 112–13
Tupuangi Formation, Chatham Islands 113, 125–6, 125
Turnbull, Mo 162
turtles 117, 120, 121, 153, 155, 168, 174
Tuarangisaurus keyesi (mosasaur) 119

Ur 29, 29–30, 58
uranium 30–1, 36, 46, 60–3, 77, 86, 126
Urewera Ranges 47, 112, 121
Ussher, Bishop 58

Venus, transit of 22–3
vertebrates 118–19
    amphibians 101, 107, 116
    birds see birds
    dinosaurs see dinosaurs
    fish see fish
    mammals see mammals
    penguins 155, 155, 168, 174–5
    reptiles see reptiles
    turtles 117, 120, 121, 153, 155, 168, 174
    whales 155, 168, 174–5, 183, 184
Victoria Range, West Coast 46
Vine, Fred 31
volcanoes and eruptions 38–55, 112–13, 187–94, 201, 207–10, 222–3
    Akaroa Volcano 149, 202–3
    Auckland volcanoes 42–3, 42, 43, 189, 222–3
    Chatham Volcano 148–9
    Dunedin Volcano 45, 148, 190, 194
    Kaipara Volcano 189
    Lyttelton Volcano 148, 194
    Mangere Volcano, Chatham Islands 125, 194, 210
    Median Batholith 84–91
    Manukau Volcano 189
    Mt St Helen's eruption 55
    Mt Somers 112–13
    Ngauruhoe 45, 54–5
    Northland Volcanic Arc 187–8
    Oruanui Eruption 48–9
    Ruapehu 48, 52–3, 54, 191–2, 223
    subduction 113, 178–9, 186, 191, 223
    Taranaki (Mt Egmont) 19, 223
    Tarawera 44, 45, 55, 223
    Taupo Eruption 50
    Taupo Volcanic Zone see Taupo Volcanic Zone
    Tongariro 19, 45, 54–5, 191–2, 223
    Tapuae-o-Uenuku 113
    Waitakere Volcano 191

Waitemata Group 189
White Island 54, 55, 179, 223
von Alsdorf, Baron 136
von Haast, Julius 162
von Hochstetter, Ferdinand 21, 40, 97, 103, 162

Waianakarua Stream, north Otago 154
Waihao River, south Canterbury 156
Waihere Bay, Chatham Islands 113, 124, 125
Waikato 191
Waikato River 49, 222–3
Waipapa terrane (Northland) 92, 94, 98, 99, 101, 103, 105, 106, 108
Waipara River, north Canterbury 154
Waipounamu Erosion Surface 148
Wairakei thermal field 47–8
Wairaki Hills, Southland 94, 95, 97, 101, 105, 111
Wairaki River 94
Wairarapa area 111, 205, 215
Wairarapa Earthquake 132, 135, 136, 141, 205
Wairarapa Fault 131, 132, 133–6, 134–5, 141, 205, 222
Wairoa 137
Wairoa River 94, 99
Waiparaconus zelandicus 155, 155
Waitahu River 71, 78, 80
Waitahu Valley 82
Waitakere Ranges 189–91
Waitakere Volcano 191
Waitemata Group 189–91
Waitomo Caves 173
Waitotara area 218
Wallis, Captain Samuel 22–3
Wanganui Basin 197, 202–3
Wangaloa Coast 154
Wangapeka Valley 82
Ward Beach, Marlborough 154–5
Waterhouse, Bruce 95, 96, 99
wave-cutting 103, 123, 147, 153, 190, 194
Weld, Frederick 141, 162
Wellington 191
Wellington Harbour 19, 136–7
Wellington Fault 131, 131, 136–7, 145
Wellman, Harold 112, 138, 139–40
West Coast 46, 71, 84, 112, 126, 127, 139, 140, 169–71, 203–4
    Punakaiki (Pancake Rocks) 168–9, 169–71
Westermann, Gerd 108
Western province, Gondwanaland 74–91, 76
    Takaka and Buller terranes 75–8, 76, 80, 82
Westland 46, 47, 80, 148, 204
wetas 111
whales 155, 168, 174–5, 183, 184
Whakamaru Ingnimbrite 54, 228–9
Whangaroa Bay, Northland 94, 98–9, 100, 101, 105
Whangaroa Harbour 188
White Island 54, 55, 179, 223
Whitebait 212
Wiffen, Joan 118, 119–22, 121
Wilson, Colin 49
Wilson, Graeme 104, 122
wind sculptures (ventifacts) 218–9
wood fossils 38, 97, 111, 113, 151, 153, 161
Wooded Peak Limestone, Nelson 96
Worthy, Trevor 120, 120, 184, 185

Yoshiaki Aita 98–9, 100, 101
Yucatan Peninsula, Mexico 128

Zealandia 16–17, 22, 24, 37, 71–3, 72, 75, 85, 164
    animals, plants, marine life 116–19, 147–57
    dinosaurs and reptiles 116–29, 119
    flat topography 146–8, 147, 165, 166–7
    impact of tectonic forces see earthquakes
    marine fossil record 153–7
        Eocene age 155, 156–9
        Oligocene age 153, 160–75
        Paleocene age 154–5
    sedimentary rocks 150, 160–75, 226
        limestone, Late Oligocene–Early Miocene 168–75
    separation from Gondwanaland 37, 47, 84, 86, 112, 117, 132, 167, 180, 226
    sinking 129, 160–75, 226–7
        land area 164, 165, 166–8, 175
        length of immersion 168
zircon 46, 60–3, 61, 70, 103

Acknowledgements

This book was the brainchild of Geoff Walker who, as publisher of Penguin New Zealand, was inspired by *In Search of Ancient Oregon* written by Ellen Morris Bishop, whom I met at a paleontology conference in Chicago in 1992. Small world! She was a leader of a post-conference field excursion along with George Stanley Jnr. (University of Montana), who is a world authority on Triassic corals.

Their excursion was mind-blowing. We started off in Wyoming looking at marine Triassic rocks on the original Canadian Shield, and then headed west through Idaho, Montana and Oregon looking at Triassic rocks that have subsequently been accreted to the western margin of North America. Plate tectonics can be blamed for this! We were looking at the mirror image of what has happened in New Zealand. Whereas Oregon is a product of tectonic processes on the north-east margin of the Pacific Plate, New Zealand is the product of processes on the south-west margin of the Pacific Plate. However, for all that we share a similar broad history relating to the Panthalassa Ocean, our histories are very different in detail.

Were it not for Ellen and Geoff, this book may not have been conceived. I for one am delighted that it was. It has given me a unique opportunity to write down in my own words how I describe the major concepts of geology and the techniques and knowledge used by geologists in determining what has happened to New Zealand the land mass. I have been communicating in this way with the public since I was appointed geologist at Te Papa. I started on the day it opened in early 1998, but I have been a professional geologist for much longer, commencing in 1978. I was lucky to be offered a position as paleontologist with the New Zealand Geological Survey, replacing Sir Charles Fleming who had retired in 1977.

As I write, I am in a lodge owned by Ian and Moana King at the end of Maipito Road near Waitangi on Chatham Island. I have the place to myself. Outside is thick fog. I am in the Chatham Islands as leader of a research team. Six of my party will be sitting in Robert Holmes cabin down at The Horns, sipping cups of coffee and waiting for the fog to lift. The Horns are remote and located in the south-west corner of Chatham Island.

My team is on a quest that will transform our understanding of the Chatham Islands. They are mapping and sampling a newly discovered limestone formation that extends more than 200 metres above sea level yet is thought to be less than 2 million years old on the basis of fossil scallops.

If these facts can be verified, then we can be certain that the Chathams have been uplifted out of the sea by some tectonic process at a rate of about 1 metre every 10,000 years. The highest point in the Chathams today is less than 300 metres. We can say with some confidence, then, that there has been land in the Chathams for about 2 million years. This means that somehow all terrestrial life on the Chathams has got there and become established within a period of only 2 million years. The implications for determining rates of dispersal, colonisation, adaptation and evolution are profound. How life got to the Chathams is a matter of fascinating speculation, but get there it did. The geology cannot be denied. May the fog lift quickly!

Much of this book was written while in the Chatham Islands. I thank my many colleagues and friends who made space for me within a busy research schedule. In particular I thank fellow geologists John Begg, Chuck Landis, Bob Carter, Alan Beu, Kat Holt, Jeremy Titjen and Chris Consoli, and the generous indulgence of Chatham Islanders Moana and Ian King, Bill and Kaye Carter, George and Ada Hough, Robert and Jan Holmes, Val and Lois Croon, and Alison Davis.

Numerous people at GNS Science have contributed in some way to the making of this book and they include: Chris Adams, Stephen Bannister, David Barrell, Julia Becker, John Begg, Kelvin Berryman, Alan Beu, Hugh Bibby, Peter Blattner, Hannah Brackley, Greg Browne, John Callan, Andrew Carman, Philip Carthew, Lyn Clayton, Ursula Cochran, Richard Cook, Maureen Coomer,

Roger Cooper, Simon Cox, James Crampton, Jan Cranston, Erica Crouch, Martin Crundwell, Mark Cunningham, Des Darby, Bryan Davy, Fred Davey, Cornel de Ronde, Gaye Downes, Donna Eberhart-Phillips, Steve Edbrooke, Robin Falconer, Brad Field, Kevin Faure, Jane Forsyth, Rob Funnell, Brett Gillies, Phil Glassey, Ken Gledhill, Ian Graham, Andrew Gray, Mary Hawkins, David Heron, Rick Herzer, Mark Hodgson, Chris Hollis, Carolyn Hume, Mike Isaac, Richard Jongens, David Johnston, Mike Johnston, Craig Jones, Liz Kennedy, Peter King, Vera Lane, Robert Langridge, Julie Lee, Graham Leonard, Margaret Low, Biljana Lucovic, Karen MacPherson, Alex Malahoff, Vernon Manville, Eileen McSaveney, Mauri McSaveney, Dallas Mildenhall, Hugh Morgans, Nick Mortimer, Bruce Mountain, Tim Naish, Neville Orr, Nick Perrin, Ian Raine, Mark Rattenbury, Martin Reyners, Russell Robinson, Karyne Rogers, Michael Rosenberg, Wendy Saunders, Heidi Schlumpf, Brad Scott, Steve Sherburn, Poul Schiøler, John Simes, David Skinner, Warwick Smith, Rodger Sparks, Ian Speden, Vaughan Stagpoole, Wendy St George, Graeme Stevens, Mathew Stott, Percy Strong, Pat Suggate, Rupert Sutherland, Richard Sykes, Suzanne Toulmin, Dougal Townsend, Roger Tremain, Noel Trustrum, Andy Tulloch, Ian Turnbull, Chris Uruski, Russ van Dissen, Mario Volk, Terry Webb, Kate Whitley, Roger Williams, Graeme Wilson, Ray Wood, Janice Wright and Marcus Vandergoes.

We have drawn heavily upon the magnificent photography of Lloyd Homer and acknowledge the value of his talent and work preserved within GNS Science.

We are especially indebted to Margaret Low for organising the photographs, Andrew Gray for expertly crafting the diagrams and Biljana Lucovic for arranging the small geological maps. Ian Graham carefully read an entire draft of the book for which we are especially grateful. His comments have saved us from many small embarrassments.

A number of people from outside GNS Science have contributed in some way or are featured in the book. We thank them for their willing assistance and participation. They include: Yoshiaki Aita, Emma Best, Neil Campbell, Rosemary Campbell, Chris Consoli, Louise Cotterall, Tony Edwards, Ewan Fordyce, Dennis Gordon, Jack Grant-Mackie, Andrew Grebneff, Tony Harris, Brendan Hayes, Rie Hori, James Jackson, Chuck Landis, Daphne Lee, Sue Maxwell, Susan Mclaurin, Simon Nathan, Adrian Paterson, John Rogers, Malcolm Simpson, Atsushi Takemura, Steve Trewick, Vivi Vajda, Bruce Waterhouse, Joan Wiffen and Trevor Worthy. We also acknowledge the support of the following institutions: Alexander Turnbull Library, Auckland Museum, DOC, EQC, Puke Ariki Museum, the Royal Society of New Zealand, Te Papa, the University of Auckland and the University of Otago.

Permission to use photographs of Rosa Ellingham and Niamh Campbell was granted by their parents Sharyn Hume and Sam Ellingham, and Hamish and Dinah Campbell.

Lastly, we thank our families for their constant if unwitting indulgence in the development of this project: Dinah, James, Saskia, Niamh and Riley Campbell; and Adele, Sam, Elinor and Isla Hutching.

Hamish Campbell, 2007

I have been fortunate enough to get to know a number of geologists over the years; their commitment and dedication to their profession have been a source of inspiration. I especially want to thank John Begg for his unfailing good advice and Hamish Campbell for his infectious enthusiasm. Both have helped make the subject a lot more understandable. Geoff Walker's perseverance has seen the project to fruition and Louise Armstrong's organisational skills have ensured the whole work came together. Thanks once again to Adele for tolerating the writer in her midst.

Gerard Hutching, 2007

The New Zealand Geological Timescale (as established in 2005)

In geology, time is formally subdivided into eras, periods, epochs and ages or stages. This is referred to as the Global Geochronological Scale, presented here. E=Early, M=Middle and L=Late. The black and white column on the left-hand side is the Geomagnetic Polarity Timescale. Black represents times of Normal polarity and white, times of Reverse polarity. All numbers on the chart are expressed in millions of years (Ma), and are based on radiometric dating.

Note that the Eras are not shown on this chart. These include the Paleozoic (Cambrian to Permian periods; 542–251 million years), Mesozoic (Triassic, Jurassic and Cretaceous periods; 251–65 million years) and Cenozoic (Paleogene and Neogene; 65–0 million years). The terms 'Tertiary' and 'Quaternary' are old names that have been replaced by the terms 'Cenozoic' and 'Pleistocene' (but are not exactly equivalent to them).

In New Zealand, geologists have established a local timescale expressed in terms of 'stages', which, for mapping purposes, are grouped into 'series'.

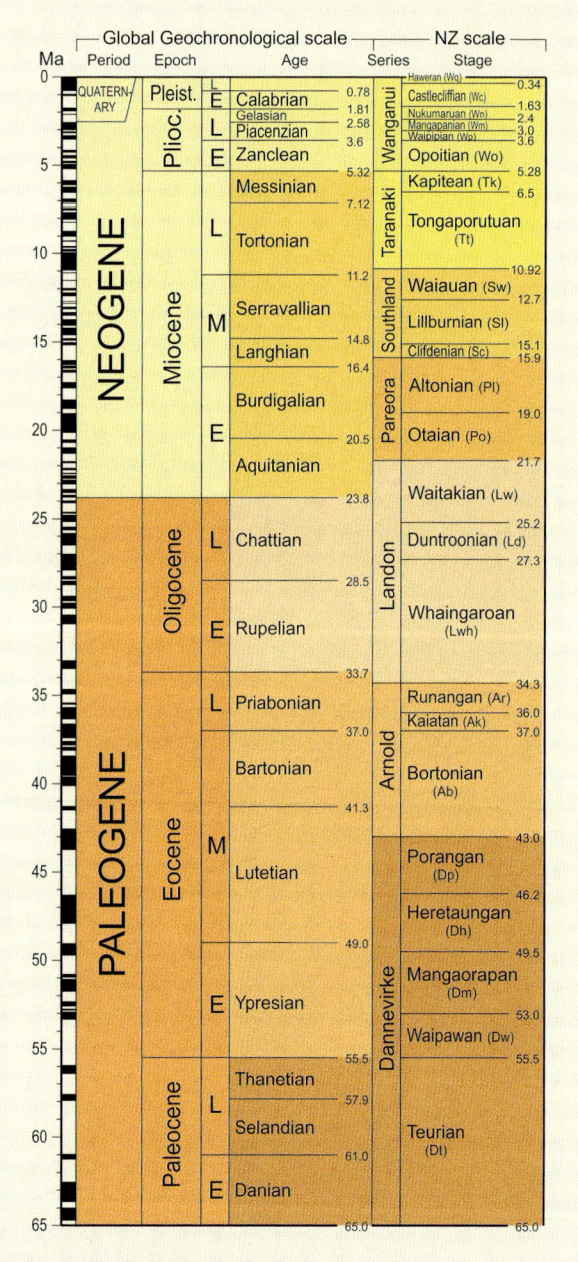

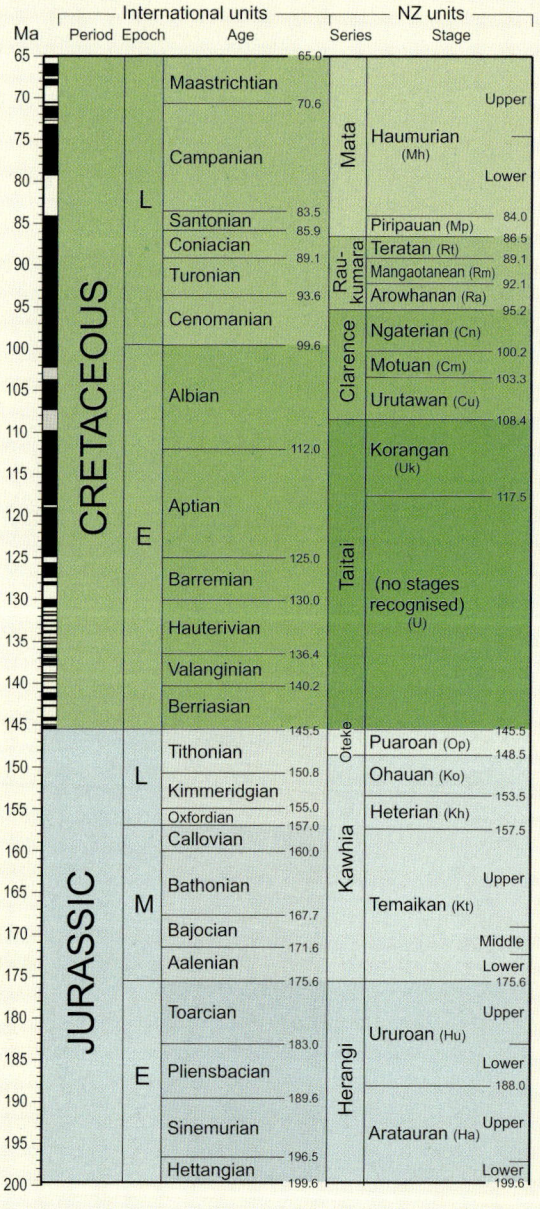

Ma	Period	Epoch	Age	NZ Series	NZ Stage
0	QUATERNARY	Pleist.	E Calabrian 0.78 / 1.81 Gelasian 2.58	Wanganui	Haweran (Wq) 0.34 / Castlecliffian (Wc) 1.63 / Nukumaruan (Wn) 2.4 / Mangapanian (Wm) 3.0 / Waipipian (Wp) 3.6
		Plioc.	L Piacenzian 3.6 / E Zanclean 5.32		Opoitian (Wo) 5.28
5	NEOGENE	Miocene	L Messinian 7.12 / Tortonian 11.2	Taranaki	Kapitean (Tk) 6.5 / Tongaporutuan (Tt)
10			M Serravallian 12.7 / Langhian 14.8 16.4	Southland	Waiauan (Sw) 10.92 / Lillburnian (Sl) / Clifdenian (Sc) 15.1 / 15.9
15			E Burdigalian / Aquitanian 20.5 / 23.8	Pareora	Altonian (Pl) 19.0 / Otaian (Po) 21.7
20					
25	PALEOGENE	Oligocene	L Chattian 28.5	Landon	Waitakian (Lw) 25.2 / Duntroonian (Ld) 27.3
30			E Rupelian 33.7		Whaingaroan (Lwh)
35		Eocene	L Priabonian 37.0	Arnold	Runangan (Ar) 34.3 / Kaiatan (Ak) 36.0 / 37.0
40			M Bartonian 41.3 / Lutetian 49.0		Bortonian (Ab) 43.0 / Porangan (Dp) 46.2 / Heretaungan (Dh) 49.5
45				Dannevirke	
50			E Ypresian 55.5		Mangaorapan (Dm) 53.0 / Waipawan (Dw) 55.5
55		Paleocene	L Thanetian 57.9 / Selandian 61.0		Teurian (Dt)
60			E Danian 65.0		65.0

Ma	Period	Epoch	Age	NZ Series	NZ Stage
65	CRETACEOUS	L	Maastrichtian 70.6	Mata	Haumurian (Mh) Upper / Lower
70			Campanian 83.5		
80			Santonian 85.9 / Coniacian 89.1	Raukumara	Piripauan (Mp) 84.0 / Teratan (Rt) 86.5 / Mangaotanean (Rm) 89.1 / Arowhanan (Ra) 92.1
90			Turonian 93.6 / Cenomanian 99.6		95.2
100		E	Albian 112.0	Clarence	Ngaterian (Cn) 100.2 / Motuan (Cm) 103.3 / Urutawan (Cu) 108.4
110			Aptian 125.0	Taitai	Korangan (Uk) 117.5
120			Barremian 130.0 / Hauterivian 136.4		(no stages recognised) (U)
130			Valanginian 140.2 / Berriasian 145.5		
145	JURASSIC	L	Tithonian 150.8 / Kimmeridgian 155.0 / Oxfordian 157.0	Oteke	Puaroan (Op) 145.5 / Ohauan (Ko) 148.5 / Heterian (Kh) 153.5 / 157.5
155		M	Callovian 160.0 / Bathonian 167.7 / Bajocian 171.6 / Aalenian 175.6	Kawhia	Temaikan (Kt) Upper / Middle / Lower 175.6
175		E	Toarcian 183.0 / Pliensbachian 189.6	Herangi	Ururoan (Hu) Upper / Lower 188.0
190			Sinemurian 196.5 / Hettangian 199.6		Aratauran (Ha) Upper / Lower 199.6

The New Zealand Geological Timescale

Ma	International units — Period / Epoch / Age	NZ units — Series / Stage
200	TRIASSIC — Rhaetian 199.6 / 203.6	Balfour — Otapirian (Bo) 199.6 / 204.6
210	Norian	Warepan (Bw) 212.0 / Otamitan (Bm) 217.0
220	L Norian	Oretian (Br)
230	Carnian 227.0	Gore — Kaihikuan (Gk) 227.5
240	M Ladinian 237.0 / Anisian 241.0 / 245.0	Etalian (Ge) 238.5 / Malakovian (Gm) 244.5 / 245.5
250	E Olenekian 249.7 / Induan 251.0	'Dur-ville' — Nelsonian (Gn) / Makarewan (YDm) / Waiitian (YDw) 250.4
260	PERMIAN — L Changhsingian 253.8 / Wuchiapingian 260.4	'Puruhauan' (YDp)
270	M Capitanian 265.8 / Wordian 268.0 / Roadian 270.6	Aparima — Flettian (YAf) 266.5 / Barettian (YAr) 273.0
280	Kungurian 275.6 / E Artinskian 284.4	Mangapirian (YAm) 280.0 / Telfordian (YAt) 283.0
290	Sakmarian 294.6	No stages recognised (Ypt)
300	Asselian 299.0	299.0
300	CARBONIFEROUS — Pennsylvanian — Gzhelian 303.9 / Kasimovian 306.5	
310	Moscovian 311.7	
320	Bashkirian 318.1	
320	Mississippian — Serpukhovian 326.4	No stages recognised (F)
330	Visean	
340	345.3	
350	Tournasian	
360	359.2	359.2
370	DEVONIAN — L Famennian 374.5	Upper — Famennian (JU) 374.5
380	Frasnian 385.3	Frasnian (JU) 385.3
390	M Givetian 391.8 / Eifelian 397.5	Middle — Givetian (JM) 391.8 / Eifelian (JM) 397.5
400	E Emsian 407.0	Lower — Emsian (Jem) 407.0
410	Pragian 411.2 / Lochkovian 417.2	Pragian (Jpr) 411.2 / Lochkovian (Jlo) 417.2

Ma	International units — Period / Epoch / Age/Stage	NZ units — Series / Stage
420	SILURIAN — Pridol 417.2 / Lud. Ludfordian 419.7 / Gorstian 422.0 / Wen. Homerian 423.5 / Sheinwoodian 426.2 / 428.4	Pridol 417.2 / Lud. Ludfordian (Elu) 419.7 / Gorstian (Elu) 422.0 / Wen. Homerian (Ewe) 423.5 / Sheinwoodian 426.2 / 428.4
430	Llandovery — Telychian 435.9	Llandovery — Telychian (Ela) 435.9
440	Aeronian 439.7 / Rhuddanian 443.2	Aeronian (Ela) 439.7 / Rhuddanian (Ela) 443.2
	Hirnantian 445.1	Upper — Bolindian (Vbo)
450	ORDOVICIAN — L Stage 6 456.1	Eastonian (Vea) 449.7 / 456.1
	Stage 5 460.5	Gisbornian (Vgi) 460.5
460	M Darriwilian 468.1	Middle — Darriwilian (Vda) 468.1 / Yapeenian (Vya) 468.9 / Castlemain. (Vca) 472.0 / Chewtonian (Vch) 473.9
470	Stage 3 472.0	
	Stage 2 479.2	Bendigonian (Vbe) 476.8
480	E Tremadocian	Lower — Lancefieldian (Vla)
490	490.0	pre-Lancefieldian 488.7 / 491
	CAMBRIAN — Furongian — Stage 6 491.5	Furongian — Paintonian (Xpa) 491.5 / 494
	Paibian	Iverian (Xiv) 498.5
500	Idamean (Xid) 501	501
	M Stage 4 501 / Stage 3 505 / Stage 2 507	Middle — Mindyallan (Xmi) 503 / Boomerangian (Xbo) 504 / Undillan (Xun) 505 / Floran (Xfl) 507
510	Stage 1 510	Ordian/lower Templeton (Xor) 513
	E No stages recognised	Lower — No stages recognised (XL)
	542	542

© Institute of Geological and Nuclear Sciences Ltd. 2005